Green Chemistry for Environmental Sustainability - Prevention-Assurance-Sustainability (P-A-S) Approach

Edited by Kinjal J. Shah

Published in London, United Kingdom

Green Chemistry for Environmental Sustainability - Prevention-Assurance-Sustainability (P-A-S) Approach
http://dx.doi.org/10.5772/intechopen.111106
Edited by Kinjal J. Shah

Contributors
Angrui Jiang, Jingwei Li, Kinjal J. Shah, Zhaoyang You, Jessica R.P. Oliveira, Giane G. Lenzi, Ntalane Sello Seroka, Lindiwe Khotseng, Keyla Soto Hidalgo, Rafael Álvarez-Chimal, Jesús Ángel Arenas-Alatorre, Faisal Nawaz, Irum Jamil, Bushra Shakoor, Anja Verbič, Gregor Primc, Martin Šala, Marija Gorjanc, Yi Yang, Suvojit Maity, Soham Sarkar, Kakoli Dutta, Tina De, Ruchira Mukherjee

First published in London, United Kingdom, 2024 by IntechOpen
IntechOpen is the global imprint of INTECHOPEN LIMITED, registered in England and Wales, registration number: 11086078, 5 Princes Gate Court, London, SW7 2QJ, United Kingdom

British Library Cataloguing-in-Publication Data
A catalogue record for this book is available from the British Library

Additional hard and PDF copies can be obtained from orders@intechopen.com

Green Chemistry for Environmental Sustainability - Prevention-Assurance-Sustainability (P-A-S) Approach
Edited by Kinjal J. Shah
p. cm.
Print ISBN 978-1-83769-389-4
Online ISBN 978-1-83769-388-7
eBook (PDF) ISBN 978-1-83769-390-0

For EU product safety concerns:
IN TECH d.o.o., Prolaz Marije Krucifikse Kozulić 3, 51000 Rijeka, Croatia,
info@intechopen.com or visit our website at intechopen.com.

Meet the editor

Dr. Kinjal J. Shah works as an associate professor at the College of Urban Construction, Nanjing Tech University, China. He is a visiting associate professor at the National Taiwan University of Science and Technology and a visiting advisor at the Climate Change Department, Government of Gujarat, India. He started his research career in 2009 at the Shah Schulman Center for Surface Science and Nanotechnology (SSCSSNT), DDU, India. During his 14-year research career, he has published more than fifty-seven journal articles, two books, and nine book chapters. He received the "Invited Lecturer from Young Scientists" award from the Society of Polymer Science, Japan (68th SPSJ annual meeting) in 2019. One of his research papers was nominated for an ENI Award in the 'Advance Environmental Solutions" category in 2019. His research focuses on the synthesis of functional materials for environmental applications, implementing green chemistry to control pollution, and physicochemical aspects of pollutants.

Contents

Preface

Environmental sustainability is a word of great importance today, as our planet's natural environment has been significantly affected by urbanization and industrialization. That is why the United Nations has set itself seventeen important goals intended to improve not only environmental but also social and economic sustainability. As responsible chemists and engineers, our greatest responsibility is to participate in the global movement to reduce global environmental impacts.

Green chemistry is a tool involving twelve principles developed by Paul Anastas and John Warner. They have moved chemistry in a more environmentally friendly direction by (1) generating "zero" waste at the source (#1), holding atoms accountable during the process (#2), and resorting to less hazardous (#3), safer (#4), solvent-free (#5), and more energy-efficient (#6) chemical synthesis routes. The principle of green chemistry also includes limiting the use of fossil materials through the use of renewable raw materials (#7), the production of materials with reduced derivatives (#8) and the use of green catalysts (#9) to avoid environmental pollution during processing. This also includes designing products for degradability (#10) and analyzing the product and process in real-time (#11) to prevent accidents (#12).

What these principles of green chemistry have in common is the potential to achieve environmental sustainability. However, there was an urgent need to prioritize the actions and therefore these principles were divided into three strategies: pollution and accident **P**revention, safety and security **A**ssurance, and energy and resource **S**ustainability. By integrating the last word of each three strategies, the Prevention-Assurance-Sustainability (PAS) strategies were developed. The basic idea was that chemists and engineers should implement these newly developed strategies into their newly developed products or processes to create a sustainable environment for the next generation.

Therefore, there is a need for books that enable integration and innovation in the theory and practice of green chemistry and provide researchers or engineers with guidance on how to develop new materials or improve existing technologies by integrating green principles into their research. This book deals with a profound solution to the existing environmental problem of sustainability.

The book consists of nine detailed chapters by carefully selected authors from experts in their fields, presenting the multidimensional side of green chemistry for environmental sustainability. These ten chapters are divided into **four** sections, with the first section containing one chapter and the other three sections containing three chapters each.

Section 1: "Theory and Practice of Green Chemistry Principles"

This section contains one chapter. **Chapter 1** is a perspective chapter that summarizes the principles and applications of all twelve principles of green chemistry as applied to control air, water, and soil pollution, along with the associated regulatory, institutional, technological, financial, and public barriers. To overcome related barriers, PAS strategies as well as the role of innovation technology and integration management have been described in detail. This section is the basic theoretical section for the principles of green chemistry.

Section 2: "Strategy P: Pollution and Accident Prevention"

Strategy P (Prevention) is the combination of the principles of waste prevention (#1), less hazardous chemical synthesis (#3), reduced derivative (#8) and design for degradation (#10). The focus of Chapters 2, 3, and 4 is on the prevention strategy. **Chapter 2** provides a good overview of the environmentally friendly and sustainable synthesis of nanomaterials, including the minimization of waste and toxic emissions. The chapter provides the bibliometric data for the green synthesis of nanoparticles provided, including the advantages and limitations of implementing such techniques. **Chapter 3** focuses on the development of greener, more economical, and environmentally friendly silica synthesis from biomass waste. This chapter provides good guidance for prevention strategies and resource sustainability. **Chapter 4** focuses on the role of mangroves and zero-valent iron nanoparticles (nZVI) for cadmium (Cd) remediation in contaminated areas. All three chapters in this section provided theoretical and experimental information to support pollution and accident prevention strategies to support environmental sustainability.

Section 3: "Strategy A: Safety and Security Assurance"

Strategy A (Assurance) is the combination of the principles of safer chemicals (#4), catalysis (#9), real- time analysis of pollutants (#11), and accident prevention (#12). Chapters 5, 6, and 7 examine assurance strategies. **Chapter 5** discusses how to provide for safer chemical synthesis of nanoparticles through biological approaches. It presents a good correlation of the biological synthesis approach with the individual principles of green chemistry. **Chapter 6** focuses on the green synthesis of Cu, Ag, and Au nanoparticles specifically for medicinal chemistry. It provides detailed information on conventional synthesis and safer chemical synthesis for nanoparticles as well as their challenges and future prospects. Meanwhile, **Chapter 7** emphasizes the development of novel and environmentally friendly zinc oxide synthesis techniques using plant extracts and gaseous plasma technology to improve the ultraviolet protective and hydrophobic properties of textile fabrics.

Section 4: "Strategy S: Energy and Resource Sustainability"

Strategy S (Sustainability) is the combination of the principles of atom economy (#2), safer solvent and auxiliaries (#5), energy efficiency (#6), and renewable feedstocks (#7). Chapters 8 and 9 mainly deal with sustainability strategies. **Chapter 8** reviews various types of green energy and their benefits to energy sustainability. **Chapter 9** is an industrial overview that describes the role of green chemistry in minimal protection of solid-state peptide synthesis. It describes how it behaves as an environmentally friendly synthesis route and increases production productivity.

This book provides comprehensive information about green chemistry for environmental sustainability, and we are confident that it will serve as an indispensable reference for engineers, scientists, and researchers in their learning about green chemistry practices for environmental sustainability. I am very grateful to all authors for their willingness to provide book chapters and collaborate with me to create awareness about *Green Chemistry for Environmental Sustainability – Prevention-Assurance-Sustainability (P-A-S) Approach*.

I would like to take this opportunity to express my sincere gratitude to Prof. Pen-Chi Chiang of National Taiwan University and Prof. Zhaoyang You and Prof. Yongjun Sun of Nanjing Tech University for their constant motivation and guidance. I also thank the group members of my KJS-laboratory for their assistance in preparing the book chapters and their participation in the book design. My special thanks go to Publishing Process Manager Mr. Tonci Lucic at IntechOpen for his support from the proposal stage to the final editing of the book. A big thank you to IntechOpen for providing this wonderful platform and supporting us throughout the entire process. I would also like to thank all my students studying sustainable green chemistry and the green chemistry for environmental sustainability course, as well as all my friends and family members for their best wishes.

Kinjal J. Shah
Associate Professor,
College of Urban Construction,
Nanjing Tech University,
Nanjing, China

Section 1

Theory and Practice of Green Chemistry Principles

Chapter 1

Perspective Chapter: Implementing Green Chemistry Principles for Pollution Control to Achieve Environmental Sustainability – A Review

Angrui Jiang, Jingwei Li, Kinjal J. Shah and Zhaoyang You

Abstract

Green chemistry is an emerging field in which 12 principles of green chemistry are put into practice to achieve a pollutant-free environment (air, water and soil). Simultaneously, when implemented in industrial practice, government policy, the practice of daily life, and the education system around the world, these 12 principles can play an important role in environmental, economic, and social benefits. This overview sheds light on the principle and its applicability based on systematic surveys. Additionally, this review identifies related barriers to GC implementation, such as: regulatory, institutional, financial, technological and public barriers to achieve the goal of a pollution free product. A three-point strategy (so-called PAS strategy, i.e. pollution and accidents prevention, safety and security assurance, and energy and resource sustainability) was explored to overcome barriers. In addition, the role of innovation technology and integration management in overcoming air, water and soil pollution system was discussed in detail. Finally, some of the valuable success stories based on GC implementation in controlling air, water and soil pollution were presented. This report highlights the success of implementing 12 GC principles in achieving overall environmental sustainability.

Keywords: green chemistry, strategies, pollution control, sustainability, innovation technology

1. Introduction

Green chemistry (GC) is a relatively new, emerging field that seeks to work at the molecular level to achieve sustainability through the design of chemical products and processes with the goal of (a) reducing the generation of pollution to minimize source and (b) risk to human health and the environment [1]. This definition and the concept of GC first came into focus in the 1962s in Rachel Carson's scholarly book Silent Spring, in which he described the devastating effects of certain chemicals on local ecosystems [2]. To protect the quality of the environment, the US Congress promptly passed the National Environmental Policy Act (NEPA) in 1969 [3]. With health and the environment now a priority, President Richard Nixon founded the United States in the 1970s [4]. The Environmental Protection Agency (EPA), a federal regulator whose very first major decision was to ban the use of DDT and other chemical pesticides [5]. In the 1980s, the chemical industry and EPA's primary focus was pollution cleanup, and industry and government leaders began an international dialog, with groups such as the Organization for Economic Co-operation and Development (OECD) raising problems and seeking preventive solutions. They made a series of international recommendations focused on cooperatively modifying existing chemical processes and preventing pollution. Additionally, in 1988, the office of pollution prevention and toxic substances was formed within the EPA to promote these environmental goals, and the term GC was first officially announced [6]. Therefore, GC goes one step further and creates a new reality for chemistry and engineering by challenging chemists and engineers to develop chemicals, chemical processes and commercial products without producing toxins and waste products. Meanwhile, the Pollution Prevention Act of 1990 marked a regulatory shift from pollution control to pollution prevention as the most effective strategy for these environmental problems [7]. By the mid-twentieth century, however, some of the long-term adverse effects of these advances could not be ignored, such as acid rain, global warming, declining forest health, and depletion of the Earth's ozone layer [1]. Some common chemicals are suspected of causing or being directly related to human cancer or other harmful effects on human health and the environment [7].

In order to overcome such problems, it is necessary to develop chemical processes and environmentally friendly products that prevent environmental pollution in the first place [8]. Through the practice of green chemistry, alternatives to hazardous substances can be created through the use of raw materials [9]. Chemical processes can be developed that reduce waste and reduce the need for increasingly scarce resources [8]. The worldwide implementation of GC requires the efforts of many governments, policy makers, academia, scientific societies, industry, trade organizations, small businesses, non-governmental organizations, research centers and other government agencies [1]. Today's industry leaders can do the job it takes chemists and engineers to move research and innovation from the lab to the boardroom by developing viable industrial products [10].

In recent years, international efforts in green chemistry have significantly boosted hopes of addressing environmental, health, and economic challenges. The International Union of Pure and Applied Chemistry (IUPAC) works with industry and other research organizations around the world to help solve this pollution and other environmental problems. Many successful entrepreneurial companies whose products are based on the application of green chemistry and technology have emerged as part of the pollution prevention program. When thinking about the development of GC, we must always remember the work co-authored by Paul Anastas and John C. Warner, who derived 12 principles (**Figure 1**) of GC that embodies a philosophy that led academic and industrial scientists to do so motivated to achieve the goal of environmental sustainability [11]. **Figure 1** shows the 12 principles that act as a clock and illustrate the importance of time, energy and the environment. The major goals of this chapter are to (1) to determine theory of GC to put in to real life practice, (2) to identify associated barriers for the deployment of GC for pollution control.

This review focuses on the detailed discussion of 12 principles and the barriers to their implementation. Furthermore, controlling air, water and soil pollution is the focus of this review, so this review sheds light on their implementation by providing strategies.

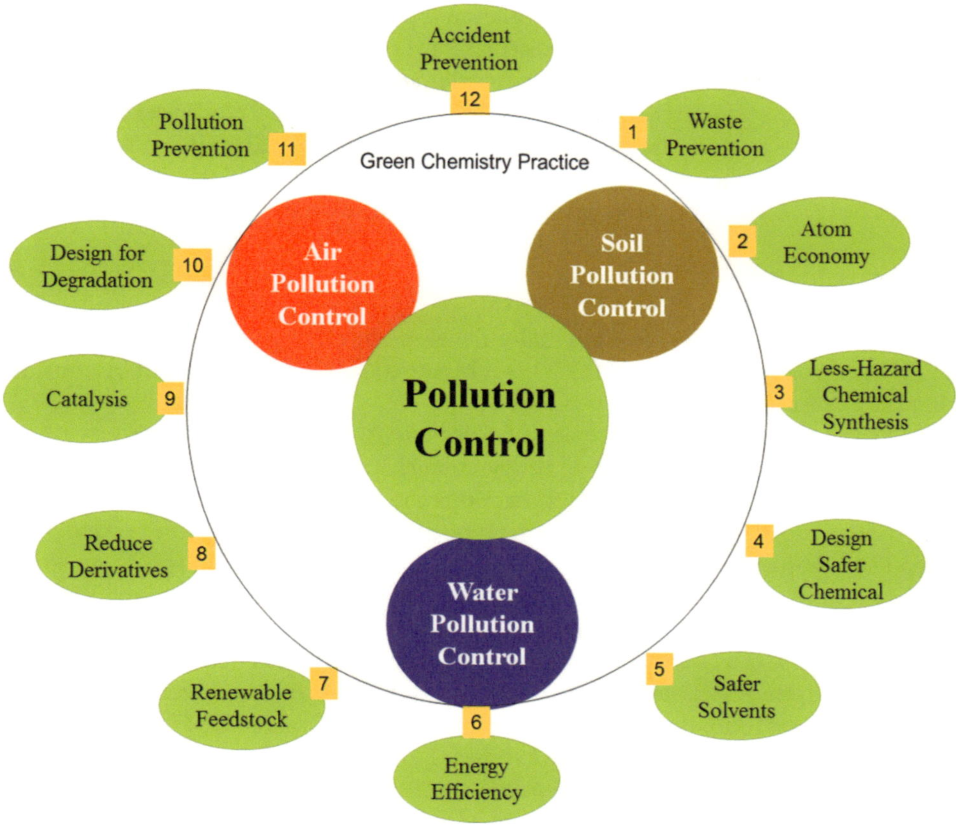

Figure 1.
The 12 principles of green chemistry to implement in pollution control.

Nevertheless, this chapter presents the role of their innovative technology in controlling each individual pollution. Nevertheless, integration management was discussed to determine the role of management in implementing the strategy. Finally, a case study on air, water and soil pollution using the green chemistry approach was discussed.

2. Twelve (12) points of green chemistry

2.1 Waste prevention

Principle: It is better to avoid waste than to treat or dispose of it after it is created. There are three sources of waste generation, for example during production, during distribution of the product and during consumption of the product [12].

Application: Waste generation in manufacturing comes primarily from sources such as incomplete reaction conversion, and waste from chemicals used in the process but not consumed in the reaction, such as solvents and suspending agents, residues, catalysts, etc. These wastes can take many forms and impacts in terms of nature, toxicity, quantity and mode of release into the environment, depending on the state. In order to control the effects, the E-factor (mass of waste/mass of products), the environmental impact factor, was introduced in 1992 [12]. It is a measure that can be

used to calculate the amount of waste generated per kilogram of product and measure the environmental impact of a manufacturing process. By providing such a factor, creative solutions to avoid inefficient industrial production have been achieved.
For ethylene oxide production, an E-factor of 5 and 0.3, respectively, was observed for the chlorohydrin process and the catalytic oxidation process [13]. This means that for every kilogram of product, 5 kg of waste was generated by the chlorination process and reduced 16-fold and 0.3 kg was generated after the catalytic oxidation process.

2.2 Atom economy

Principle: Atomic economics is a major development that goes beyond the traditional concept of percent yield [13]. This is because the yield is calculated by considering only one reactant and one product.

Application: The central principle of GC is to design processes in such a way that the greatest possible amount of all raw materials ends up in the product and a minimum of waste is produced. Most reactions can have a high percentage yield, but also produce a lot of waste products. This type of reaction has a low atom economy. When designing an environmentally friendly chemical process, both yield and atom economy must be considered. Ethylene oxide is an important raw material for the production of epoxy-functional materials, ethoxylates (surfactants), ethylene glycol ethers (solvents), etc., which are produced by the chlorohydrin process (two-step process, see **Figure 2A**). Meanwhile, in 1937, Union Carbide introduced a catalytic oxidation process (one-step process) to produce ethylene oxide (see **Figure 2B**) using silver metal as a catalyst. **Table 1** shows the atom economy calculations of both processes. As a result, it was found that the atom economy of the chlorohydrin process and the catalytic oxidation processes were 23 and 100%, respectively [13]. Silver is the most active and selective catalyst for the reaction. Today, all ethylene oxide worldwide is produced exclusively via the catalytic route.

2.3 Less hazardous chemical syntheses

Principle: Synthetic methods should be designed in such a way that the formation of substances (product or by-product) has little or no toxicity to human health and the environment.

Application: Chemical products should be designed to prevent the effect while reducing toxicity. In practice, many researchers worked to reduce solvent consumption and produce more efficient processes with higher yields and less hazardous

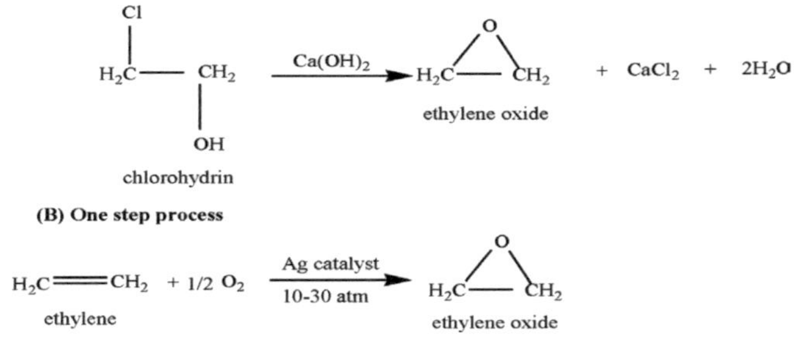

Figure 2.
Synthesis of ethylene oxide by (A) chlorohydrin process, and (B) catalytic oxidation process.

Process	Reactant		Product				Atom economy
			Utilized		Non utilized		
	Formula	Formula weight (g/mol)	Formula	Formula weight (g/mol)	Formula	Formula weight (g/mol)	
(A) Chlorohydrin process (E-factor = 5)	C_2H_4	28	C_2H_4	28			
	Cl_2	71		0	Cl_2	71	
	H_2O	18	O	16	H_2	2	
	$Ca(OH)_2$	74		0	$Ca(OH)_2$	74	
	$CH_4H_8O_3CaC_{l2}$	191	C_2H_4O	44	$H_4O_2CaCl_2$	147	44/191 = 23%
(B) Catalytic oxidation process (E-factor = 0.3)	C_2H_4	28	C_2H_4	28		0	
	½ O_2	16	O	16		0	
	C_2H_4O	44	C_2H_4O	44			44/44 = 100%

Table 1.
E-factor and atom economy of ethylene oxide produced by (A) chlorohydrin process, and (B) catalytic oxidation process.

waste/intermediate materials [14]. Building on this, the 2005 Nobel Prize in Chemistry was awarded for advances in green chemistry. To illustrate this concept, a Witting reaction invented by Georg Wittig in 1954, in which he incorporated an alkane into a molecule, was highlighted [15]. A triphenylphosphonium halide was added to a solution of n-butyllithium in diethyl ether. After ylide formation, the carbonyl compound was added and refluxed overnight. After some time, a milder base was used in place of the lithium compound to form the ylide, and N, N-dimethylformamide was used in place of diethyl ether. The Witting reaction was modified with aqueous sodium hydroxide at room temperature to aid GC in solvent-free synthesis. Here, an aromatic (any) aldehyde is added to a ylide formation by reacting benzyl-triphenylphosphonium chloride with sodium hydroxide by simply stirring at room temperature for 30 min (see **Figure 3**) [14]. The advantage of such a reaction was the replacement of the reaction solvent with water.

2.4 Safer chemicals

Principle: Chemical products should be designed to perform their desired function and effectiveness without compromising on a minimal level of toxicity.

Application: To date, most industrial processes have been developed and aligned with the end goal of producing a product and its utility, regardless of the impact on natural resources, the waste generated by the invention, and its toxicity [11]. To understand this concept, it is essential to understand the flame retardants used in the plastics industry. In general, organ halogen and organophosphorus compounds are widely used as flame retardants. In the organ halogenated compound, brominated aromatics in particular have been the most commonly used because of their low cost and efficiency. However, it exhibits less bio-accumulative properties in the environment and the associated health risks are significantly higher. More recently, bio-based flame retardants are derived from tartaric acid, a by-product of winemaking used in the polymer composites industry [16].

2.5 Safer solvents and auxiliaries

Principle: Solvents are the king of all types of reactions (mainly) to get a good conversion rate. The use of auxiliary materials (e.g. solvents, separating agents, etc.) should be superfluous as far as possible and harmless in use.

Application: In many cases reactions would not take place without solvents and/or material release agents. However, solvents are responsible for about 75% of the

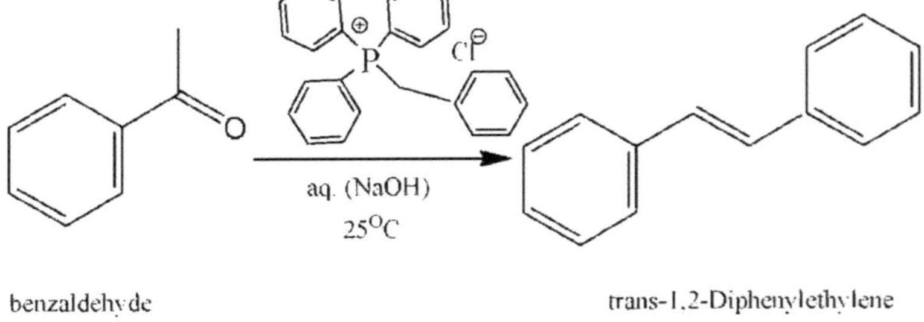

Figure 3.
Green route scheme for the wetting reaction of substituted benzaldehyde.

total environmental impact of a traditional chemical operation [17]. In addition, they lengthen the process steps by being heated, distilled, cooled, pumped, blended, distilled under vacuum and filtered, and for the most part are not recycled. To avoid such effects, in some cases microwave-assisted solid-phase reactions, green solvents, and water media were alternative ways to reduce solvent consumption [18]. Among other things, the microwave-assisted reaction is very fast, has a shorter reaction time, a higher degree of purity and a higher yield after the reaction [18]. Most importantly, no additional steps are required to obtain a pure product.

2.6 Energy efficiency

Principle: requirement of chemical processes should be recognized and minimized with regard to its ecological and economic effects.

Application: In the general case of solvent removal, chemistry laboratories need a rotary evaporator, which requires the combined use of a heat source, a vacuum pump, a rotary motor, and a condenser. Meanwhile, the energy input and associated costs are modest compared to a larger industrial scale for solvent removal. To control the energy consumption, you need to design the reaction under the ambient conditions of room temperature and atmospheric pressure [11]. In 2012, Cytec Industries Inc. received the Presidential Green Chemistry Challenge Award for innovation in energy-saving reaction conditions. In short, alumina is a raw material for aluminum, which is produced from bauxite using the Bayer process. Minerals were deposited on heat exchangers and pipes that were cleaned with sulfuric acid when the engine was shut down. To avoid this problem, Cytec developed its MAX HT Bayer Sodalite Scale Inhibitor, which contains an active polymer with silane functional groups for the Bayer process. This polymer deposited on surfaces, increased evaporation and reduced caustic soda losses, thereby reducing the use of vapor emissions from the combustion of carbon-based fuels. This process also reduces the use of sulfuric acid to clean heaters. After implementing this process at all steel mills, it was determined that the energy savings equated to approximately 7 billion pounds of CO_2 not released into the atmosphere. In addition, the use of sunlight to promote chemical reactions and the conversion of solar fuels from CO_2 offer significant potential for sustainable energy solutions [19].

2.7 Renewable feedstocks

Principle: A raw or starting material should be renewable and non-exhaustive whenever technically and economically feasible.

Application: Recently, most of our manufacturing products are derived from petroleum feedstocks or natural gas. Renewable raw materials are often made from agricultural products or come from living organisms [20]. Many products from biological sources are widely used, e.g. Algae is used in the cosmetics industry, lignin is an important waste material from the pulp and paper industry for energy production, additives are converted into DMSO as well as humic acid, starch and oil for the production of detergents, bioplastics are Synthesis for packaging, cars, etc. [21]. Among biopolymers, polylactic acid is the most commonly used biopolymer in the packaging and healthcare sectors today [22]. Fermenting starch derived from glucose in the presence of CO_2 and yeast produces succinic acid, which is used in the synthesis of a variety of chemical products such as detergents and biodegradable plastics. In recent years, CO_2 has been used as a feedstock for mineralization in the cement industry through the carbonization process to the High-Gravity process. Similarly, bio-butanol is also produced from renewable feedstock and can be used for many energy applications [23]. Composed of triglycerides, vegetable oils are widely used in foods, pharmaceuticals, cosmetics, fuels, paints and building materials. Research into the use of renewable materials is advancing rapidly and one cannot hope to provide an exclusive overview of all possibilities.

2.8 Reduce derivatives

Principle: Unnecessary derivatization should be minimized or avoided whenever possible, since steps require additional reagents and can generate waste.

Application: Synthesis is a multi-step process to produce chemical products using various solvents and reagents at elevated temperatures and/or pressure. This process requires additional purification and separation steps after each process and generates waste that gives the impression that the recovery is less than 100%. Noncovalent derivatization is an alternative to traditional synthesis, in which the properties of a starting material are modified by covalently attaching various functional groups to it without the use of solvents. Thus, no waste and no purification is required to obtain modified products. To understand this concept, it is necessary to understand the Polaroid instant imaging mechanism. Here, the solubility of hydroquinones at elevated pH by bis(N, N-dialkyl)terephthalamide modification was modified by a noncovalent derivatization method instead of the (traditional) method using base-labile covalent protecting groups [24]. It solved the problem without altering the original hydroquinone structures and minimized waste material and energy by performing the solvent-free milling, purification processes, and waste disposal associated with the traditional covalent derivatization process in one step.

2.9 Catalysis

Principle: Catalytic reagents are superior to stoichiometric reagents to increase product yield. As described in the first principle, waste avoidance can be achieved through a higher conversion rate, which can be achieved by a catalyst.

Application: One of the most important catalytic reactions of the last century is the acid-catalyzed Friedel-Craft alkylation and acylation reaction for the synthesis of alkylbenzenes, which are valuable intermediates in specialty chemistry. In this reaction, soluble Lewis acid AlCl3 or mineral HF acid was added as a catalyst and resulting in a large amount of hazardous substances (**Figure 4a**) [25]. To cope with this situation, solid acid catalysts have been developed from an economical and ecological point of view to achieve higher yields with fewer pollutants (**Figure 4b**). In this century there was the renewable catalyst or enzyme catalyst. Enzyme catalysts were used to convert the pollutant gas CO_2 into useful solar fuels (**Figure 4c**) [26]. In this way, harmful substances are converted into useful substances. Furthermore, since biocatalysis is a biomimetic approach based on natural or modified enzymes, it is also used for the conversion of biomolecules instead of petrochemicals.

2.10 Design for degradation

Principle: Chemical products should be designed so that at the end of their useful life they break down into harmless decomposition products and do not remain in the environment.

Application: There is a risk of confusion in understanding this concept, where biomaterials are not necessarily made from renewable resources, but refer to the

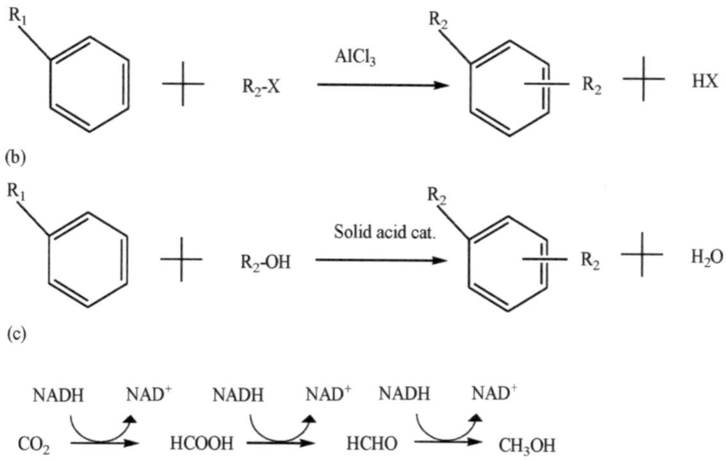

Figure 4.
Catalytic reaction such as Friedel-crafts alkylation by (a) conventional route, and (b) green route; enzymatic reaction for solar fuel production (c).

ability of materials to break down into energy and more basic components such as carbon dioxide, methane and water. Biodegradation occurs aerobically and anaerobically, with organic matter being converted into basic elements in the presence or absence of oxygen. In the last 10 years, due to the global worm effect, the demand for sustainable and renewable raw materials for personal care products has increased. Prior to this type of concept, long-chain ammonium salts such as di(hydrogenated) tallow dimethyl ammonium chloride (DHTDMAC) (**Figure 5**) were used as personal

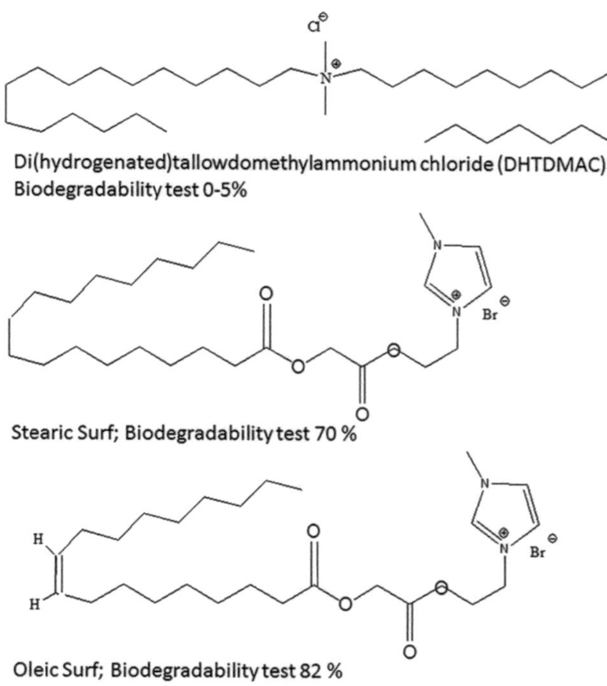

Figure 5.
Design of various biodegradable surfactants.

care products. However, it has lower biodegradability and intrinsic eco-toxicity [27]. The new salt made from natural fatty acids has higher biodegradability compared to DHTDMAC (see **Figure 5**). Made from fatty acid, the surfactant has superior physicochemical properties, is easily synthesized and over 70% biodegradable, providing a good alternative to traditional petrochemical-based surfactants.

2.11 Real time analysis of pollution

Principle: Analytical methods need to be further developed to enable real-time monitoring and control during the process before hazardous substances are formed. In most chemical processes, temperature and pressure are linked.

Application: Temperature and pressure sensors were used to monitor the reaction conditions. However, they only measure the condition of the reactor, but the composition of the mixture, the conversion ratio of the mixture and the physical properties must be measured with analytical techniques (see **Table 2**) to avoid generating waste. The product/process is safer for human health and the environment. Through this type of practice, the goal of green analytical chemistry to measure chemicals without generating waste and minimizing or eliminating the formation of by-products can be achieved [11]. The advantage of the real-time analysis process is to quickly take action to avoid accidents, energy loss and by-products that would survive additional cleaning, separation or filtration steps. In particular, materials used in the manufacture of such an instrument, such as a mercury electrode replaced with a carbon electrode and a Li-ion battery replaced with a carbon electrode battery, are also considered. After implementing such small things for production, pollution and by-products can be eliminated and the product obtained can be called a clean product.

2.12 Accident prevention

Principle: Substances and the form of a substance used in a chemical process should be selected to minimize the risk of chemical accidents, including releases, explosions and fires.

Application: When it comes to prevention, not only chemical accidents count, but also protection against damage, risks, dangers, injuries and damage. In general, accident prevention is divided into four categories: personal protection, management and control of work processes, technical control and elimination of the accident before it occurs. According to the OECD principles of good laboratory practice, the generation of high-quality and reliable test data/products is related to personal and industrial safety. Personal safety through awareness, cleanliness, safer work culture and use of personal protective equipment such as gloves, goggles, lab coat/apron, shoes and masks. The Occupational Health and Safety Administration guidelines must be followed in controlling and managing work practices to prevent accidents. Meanwhile, according to Environmental Protection Agency guidelines, materials and processes must be selected to be safer for the environment. Under the Chemical Accident Prevention and Clean Air Act Amendments of 1990, accident prevention begins with identifying the hazards, such as: exclusive, toxic, flammable or global hazards. In this way, the production or laboratory work is the first to benefit from the risk reductions. There are several serious accidents related to the chemical industry or chemical materials, resulting in significant losses to people, animals and the environment, for example. The above incidents are memorable events for everyone, especially the scientific community, reminding us to use safer alternative chemicals to avoid accidents.

Type of Basic Instruments	Type of Industries							
	Basic Inorganic	**Drug**	**Dyes intermediate**	**Fertilizer**	**Refineries and petroleum/ Polymer**	**Pesticide**	**Electroplating and heat treated**	**Oil and soap**
Gas chromatography	Salts	Drug Molecule and purity	Composition of mixture	Reaction conversion and purity	Yield and Conversion rate	Conversion rate	Toxic chemicals and metals	Spent soap lye
Raman spectroscopy	Acid/ Alkalis	Molecule	Structure of product	Molecular structure	Molecular structure and bonding	Molecular structure	Bonding of materials	Emulsion rate
Optical spectroscopy	NA	NA	NA	NA	Composition of a real mixture	NA	Coating of transparent material	Wettability and contact angle
UV/Vis spectroscopy	Inorganic pollutants	Solid pollutant	Water-soluble pollutants	Pollutants	Vinyl polymeric conversion	By-products	Bond formation rate	By-products
Infrared spectroscopy	Inorganic pollutants	By products	By products	By-products	Structural detail of product and waste	Structural detail of product and waste	Functional group of coating materials	Functional or coating groups

Table 2.
Type of analytical techniques utilized in different kinds of industries.

3. Goals and objectives

Government, policymakers, business, industry and the public now see sustainable development as a necessary goal to achieve the goals of the Paris Agreement of November 4, 2016 [28]. In the context of industrialization, the creation of cleaner products and cleaner processes plays a key role in reducing pollution minimize and to maintain and improve our atmosphere and thus our quality of life. This can be achieved through GC Development. In fact, these challenges are primarily associated with chemistry, chemical engineering and chemists. The reason the manufacturing process creates millions of tons of waste, and reducing or eliminating that waste is a critical issue facing the world today. Therefore, a challenge for chemists and their associations is to develop new products, processes and services that achieve the goal of a pollution-free environment. This can be achieved by reducing the material and energy intensity of chemical processes and products, minimizing or preventing the spread of harmful chemicals in the environment, maximizing the use of renewable resources, and extending the shelf life and recyclability of products. To achieve the goal of a pollutant-free atmosphere, the aim of the paper was to (1) identify the principle of GC and the role of GC in preventing pollution, (2) the obstacles and challenges in establishing GC in the system to identify, (3) The role played by innovative technologies in preventing air, water and soil pollution is determined. In addition, three case studies on different countries' approaches to tackling GC-related challenges were reviewed and illustrated.

4. Barriers and challenges

Basically, when we talk about the barriers, it is necessary to consider a complex set of issues associated with the implementation of GC for pollution control. To justify that barriers are categories of institutional, regulatory, technological, financial and public (cultural) support to achieve comprehensive pollution control. There are significant interactions and overlaps between these categories themselves.

4.1 Regulatory barriers and challenges

In recent years, income growth has changed according to the country's urbanization development, which has led to an increase in export and import. In addition, with the development of the transportation system, the world is becoming more and more connected, and the export and import of chemicals is increasing rapidly. Therefore, many chemical companies have established their business according to the relaxation of regulations in the country where they manufacture or sell their products. Many chemical companies use their resources for mandatory actions and harm the environment, air pollution and water instead of investing in research and development for new products and processes. In most cases, landfilling the waste is not the ultimate option, and the quality of the landfill is also questionable if it is not properly segregated. Such governmental habits and underestimations can be a barrier to new GC innovations. In addition to waste management, the government must also take care of corporate policy to avoid conflicts [29]. In some cases, the sale of one company or product interfered with the sale of another product, leading to conflict and implementation difficulties. At this point, government environmental and health policies must consider the benefits and make their choices rather than being economic or political [30]. In addition, the central government must provide adequate rights and funds to the state government, which has the right to promote GC to achieve pollution-free products, so-called green products.

4.2 Institutional barriers and challenges

There are many institutional barriers associated with GC to implement it for pollution-free production. Political issues, lack of policies, taxes, and outdated infrastructure hinder the creation of an effective GC system. Industry and academia are now responding to these challenges and implementing them on a global scale, attracting research funding and moving new, greener chemicals from the research bench to the manufacturing plant. In this implementation, a lack of awareness was observed in education in schools, universities and companies, and in management perceptions [31]. In the implementation, GC are important to bring the laboratory experiment to industrialization. Beyond these barriers, the region faces several other constraints, including a lack of coordination, transport dependency, aging infrastructure, operational limitations, and a broader public and industrial perception problem. Another major obstacle is the successful implementation of new synthesis technologies and corresponding new process technologies, especially in the pharmaceutical industry where US FDA and other documentation pose challenges over time. With such institutional and organizational obstacles, international agencies and national governments need to set up appropriate institutional structures to reduce process time, but also try to avoid disruption from frequent institutional changes. The lack of a common environmental objective across the different institutions can represent a significant obstacle to implementation.

4.3 Technological barriers and challenges

Innovation technology and management integration is the task of the coming century to overcome environmental barriers through the synthesis of green products, so that green economics becomes the driving force behind new products and processes. To implement technology in GC to achieve green production, aspects must be multidisciplinary and not specific. The reasons for this are lack of knowledge about toxicity among chemists and vice versa, the knowledge of laboratory chemists and industrial chemists differs depending on experience and finally, there is no such database of chemical lists and process lists that could help the chemist to establish a product. This is because the industry is unwilling to share information about the chemicals used in production and synthesis processes. As a result, the many substances, hazard characteristics and processes remain unknown. There are no clear design guidelines for researchers in the early discovery phases of GC as there is no communication gap between scientists and engineers. A concerted research initiative by industry, government and academia is required to develop a wide range of reliable, harmless methods available when needed. As part of this initiative, the Crystal Faraday Partnership was established in the UK, where the research forum is funded by both industry and government to promote research, education and dissemination in green chemistry and processing [32]. In this way, the last technical hurdle of the data and information gap was overcome. Still, there are many gaps in knowledge sharing that pose barriers to implementing GC and GT to achieve cleaner production.

4.4 Financial barriers and challenges

GC represents a major cultural change in the way chemicals are developed and materials and products are manufactured. This practice focuses on building a workforce of trained chemists with specializations in toxicology, GC principles, and environmental sustainability. But the investment and promotion of state institutions is not part of the mainstream. In most cases, the benefits of GC accrue directly to the public rather than to an individual company. Therefore, the design of the investment must be such that the impact on the business unit can be reduced, so that new investments for the modernization of traditional units can be easily justified in a profitable way without sacrificing efficiency and quality. In the pharmaceutical industry and for companies with large capital investments, the cost of shutting down an old facility is so high that companies do not have the resources to subsequently reinvest in new technology. In this type of industry, large reactors were used to produce fine chemicals, causing energy losses, solvent losses, and resource losses [33]. This is the type of case where a greener process could easily be deployed, but the companies are tied to the facility, US FDA approval, and a lot more paperwork. Such a situation is not limited to the pharmaceutical industry, but specialty chemicals and all kinds of chemical manufacturing industries also have to overcome such hurdles [34]. Just in the event that the company is able to decommission the old infrastructure, the company could face a potential financial drain that makes maintaining the status quo much more attractive than investing in new infrastructure.

4.5 Public support

The implementation of GC was not only related to chemists and companies, but also to the participation of the general public. While GC has made initial strides, chemists and all are misunderstood as a school curriculum and part of the environmental agenda for washing and cleaning pollutants and waste. On the other hand, due to lack of awareness and the like, they do not have much idea that this new stream is of use to them and brings many benefits [34]. In this scenario, the general public is far from understanding the benefits of GC and does not accept the costly and less effective products that are environmentally friendly. Another major obstacle arose when the public began to consider separating academia and industry, and this concept restricted only the publication of peer-reviewed articles and not industrial research. There are major obstacles in the public sphere regarding highly innovative industries where technology and demand are changing rapidly due to changing fashions and requirements. Without public support in the industry, there are significant obstacles to the implementation of GC. The state police must sensitize the public to waste management and recycling concepts in order to achieve acceptance and many successful implementations of the policy. Therefore, raising public awareness is very important, and in addition to raising public awareness, the public must also be involved in policy formulation to create visions for the future.

5. Strategies to build GC

To promote GC for non-toxic manufacturing, the Green Chemistry Commerce Council (GC3) was formed in 2005 to advance this growing field by bringing together key business leaders, governments, experts, academic researchers, and environmental and health advocates [35]. The philosophy of GC3 is that challenges in exploring, adopting and scaling GC are most effectively solved through cross-industry dialog, collaboration and supply chain partnerships [36]. Five key strategies have been developed to implement GC: enhancing market dynamics, supporting smart policies, fostering collaboration, and informing the market and tracking progress. However, this approach to address the increased GC demands requires an additional new approach, namely P-A-S (Prevention, Assurance and Sustainability) proposed by Chen et al. is suggested [1]. As such, there is a shared responsibility in implementing the 12 GC Principles that can help achieve environmental sustainability, i.e. (1) Prevention of pollution and accident; (2) Assurance of safety and security; (3) Sustainability of energy and resource. This PAS strategy was developed by integrating the first word of the each three parts. **Figure 6** shows the incorporation of 12 GC principles and international movements according to PAS strategies. It can be seen that waste prevention (#1), less hazardous chemical synthesis (#3), and reduced derivatives (#8) are core component of the prevention strategy, meanwhile design for safer chemicals (#4), real-time analysis of pollutants (#11), and accident prevention (#12) are core component of the assurance strategy; and atom economy (#2), design for energy

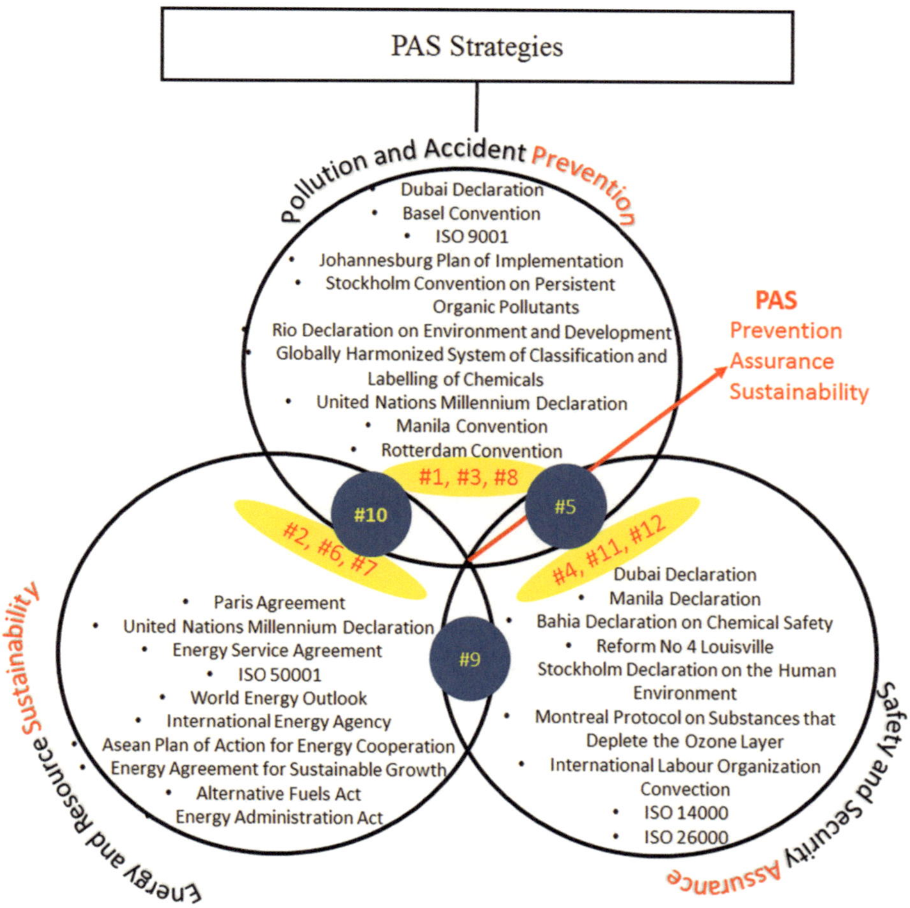

Figure 6.
PAS strategies incorporated in 12 green chemistry principles and international movements.

efficiency (#6), and renewable feedstock (#7) are core component of the sustainable strategy. Moreover, principles #5, #9 and #10, namely safer solvent and auxiliaries, catalysis and design for degradation, are hybrid components of prevention and assurance; assurance and sustainability, and sustainability and prevention, respectively. In addition, **Figure 6** shows the list of international movements categorized by PAS strategies. This new approach is a potential tool that can be used in government, industry and educational institutions around the world to mitigate global sustainability issues. After building these three strategies, three short-term goals need to be set, namely: (1) scaling GC innovation; (2) increase the importance of GC in daily life by accepting the concept; and (3) develop and adopt smart policies that can support industry, research and innovation. In that regard, innovation technology and integration management have also been brought together with these strategies to strengthen PAS strategies.

6. Innovation technologies

Innovation technology in GC offers opportunities to discover and apply new synthetic approaches using alternative starting materials; environmentally friendly reaction conditions, energy minimization and the development of less toxic and inherently safer chemicals [28]. In the nineteenth century, environmental protocols changed and the LeBlanc soda process was redesigned to be more environmentally friendly. This was the first noticeable change towards cleaner production to get an environmentally friendly product. Apart from this process, the Montreal Protocol calls for replacing the chlorofluorocarbon compound to protect the ozone layer. Waste greenhouse gases such as N2O were used in the production of phenol. In addition, tetrakis(hydroxymethyl)phosphonium phosphate is used as a low-dose agent to combat microbial growth in industrial refrigeration systems with low toxicity.

6.1 CO_2 utilization for the production of valuable fuel/chemicals (air pollution)

Harnessing the emissions of CO_2 into the atmosphere is one of the key elements in controlling global warming. There are various methods, namely direct use by microalgae, conversion to chemicals, conversion to energy, and use by solar energy as fuel. The use of CO_2 to form glucose (**Figure 7a**) and carbonization products has been observed in nature and in industry, respectively (**Figure 7b**). Through this technology, valuable chemicals were converted or used in an indirect conversion process. However, a conversion to fuel from direct CO_2 is still being examined. Inspired by natural photosynthesis, systems for the production of artificial solar fuels are now being developed and their efficiency tested. Glutamic acid has been used as one of the conversions for solar fuel (**Figure 7c**) [26]. Similarly, the enzymes FDH, AlDH, and ADH along with NADPH have been used for the conversion of solar fuels, which represent the formation of format and CO by FDH reductase. However, the conversion still takes place in an aqueous phase rather than in the air, so the solar fuel conversion system was developed (**Figure 7d**). Laser light was used in the system to initiate the photoreaction and successful stepwise conversion was achieved, for example, of formic acid from adsorbed CO_2, formaldehyde from converted formic acid, and methanol from converted formaldehyde.

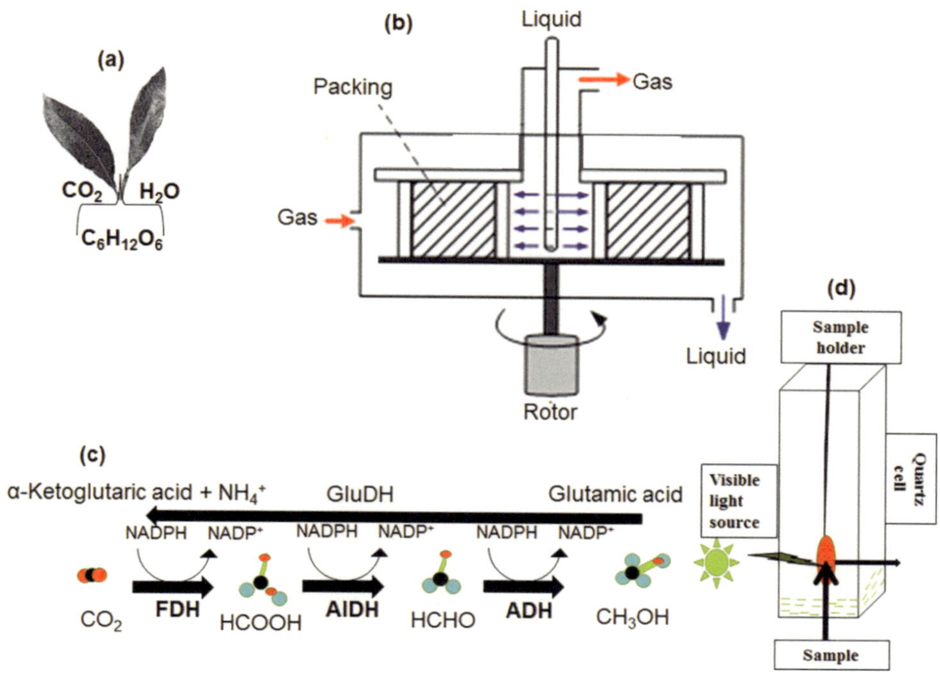

Figure 7.
CO_2 utilization in (a) nature plant, (b) high gravity rotating pack system, (c) multi-enzyme system, and (d) solar fuel mechanism.

6.2 Green solvent technology (water pollution)

Many of the industrial manufacturing and laboratory syntheses have changed their manufacturing process to a catalyst-free, solvent-free and safe synthesis. As a result, many named reactions, multicomponent reactions and the heterocyclic compounds in water and on water, are switched to catalyst-free reactions by microwave irradiation and ultrasonic irradiation. In addition, ionic liquids are materials that have attracted significant attention in the scientific community due to their biodegradability, green nature, and environmental considerations, including negligible volatility, nonflammability, and stability (thermal, chemical, and electrochemical). Other solvents such as supercritical fluids (CO_2 and water) have been used as alternative solvents due to their lower critical constant, non-toxicity, and inexpensive nature. Supercritical CO_2 is able to separate low-volatile, higher-molecular and/or more polar compounds with increasing pressure. Apart from these solvents, solvents can be derived from natural compounds and replace petroleum-based solvents. This concept won 22 Presidential Green Chemistry Awards between 1996 and 2014 for approaches that reduce the use of traditional solvents, including alternative, more environmentally friendly solvents [36], and processes that use carbon dioxide and water or eliminate the use of solvents entirely.

6.3 Greener electrochemical storage systems (batteries)

As the shell of the earth, the soil is the most important part of the earth as most human activities are closely related to soils. Stabilizing agents have been developed to promote adsorption, complexation and precipitation to stabilize heavy metal effluent.

In general, lime, zeolite, minerals, mining waste, soil, clay, and biochar have been used to control heavy metal accumulation [37]. Many green modifiers have been developed to effectively immobilize lead, zinc, copper and cadmium in contaminated soils, thereby significantly reducing the bioavailability of pollutants. Cultivated treatment wetlands, phytoremediation and bio-retention are the applicability of this type of bio-accumulated modified filler to improve soil and water quality. In the engineered wetland, saturated substrate, vegetation, and microbes were used to mimic the natural wetland. Here, green chemistry approaches were applied to improve the adsorption quality through sedimentation, plant uptake and filtration mechanism. This green wetland also has the potential to biodegrade pollutants into non-toxic forms, be adsorbed by wetland plants, adsorbed onto the adsorption surfaces, or chemically converted to be stored in the wetland matrix. This system offers advantages over traditional mechanisms in terms of cost efficiency due to locally available materials, no additional environmental concerns and is easy to build. This is the area of GC that requires the highest level of innovation to deliver the next generation of innovations in support of a sustainable environment.

7. Integrated management

GC management activities can be divided into three types: (a) cleaner laboratory practice, (b) cleaner production, and (c) sustainable consumption [29]. Clean laboratory practices are behaviors and education aimed at creating more environmentally friendly methods of laboratory work in order to avoid major environmental damage in large-scale production. To achieve this, it is necessary to discuss how the GC approach can effectively minimize hazardous waste in science by including such subjects or courses in curricula. Today, "pollutant-free production" means more environmentally friendly product manufacture in a broader sense with fewer dangerous emissions. When it comes to cleaner production, there are two important aspects to consider, namely industrial ecology and the life cycle of materials. To achieve this goal, several agencies, namely the Green Chemistry Institute (GCI) in the US, the Green Chemistry Network (GCN) in the UK, and the Green and Sustainable Chemistry Network in Japan, are working to bridge the gap between science and academia to close. The manufacturing company suggests sharing knowledge. Not only must management achieve this through cleaner manufactured products, but it can also be achieved through a greener application of the supply chain through to the end use of the product. After manufacturing the product, sustainable consumption is very important, taking into account the need for chemicals, so there is no alternative. Therefore, the safety of chemicals is less important than the need and justification for chemicals having an impact on the environment. This is how green business and cross-industry cooperation must take place. So consumption is just as important as production. All three things above are divided into three levels of management where laboratory practice, production and consumption take place respectively at micro, meso and macro levels.

8. Case studies

8.1 Newlight technologies (air pollution)

Newlight technologies developed AirCarbon, a commercial carbon capture technology that creates AirCarbon by combining methane with air. Basically, AirCarbon is a thermoplastic high-performance material. This technology works at ambient temperature, with the developed catalyst combining air with methane-based carbon to produce thermoplastic materials [38]. In general, most of the technologies available on the market capture CO_2 and produce plastic, but the production costs are significantly higher than petroleum products. To overcome this obstacle, Newlight developed a biocatalyst and method to deactivate the negative receptors on polyhydroxyalkanoate polymers and produce inexpensive and highly efficient catalysts for plastic production. This is the main polymer-producing enzyme in the biocatalyst, capable of polymerizing 9 kg of polymer per 1 kg of biocatalyst used. In this way, methanol is used, which is responsible for capturing the radiation. The carbon negative plastics have been used to manufacture many useful products such as plastic bags, cell phone cases, home accessories and other products. In addition, many leading companies have adopted the AirCarbon product as part of their packaging materials for bags and cell phone cases. By eliminating greenhouse gases, New Light Technologies won the 2016 EPA Presidential Green Chemistry Challenge Award for pollution-free manufacturing.

8.2 Pfizer Inc. (water pollution)

Zoloft, the most prescribed drug for depression, is consumed annually in the United States, valued at $43.7 billion. Pfizer pioneered production using sertraline as the active ingredient, which was synthesized in a three-step process during the original manufacturing process and reduced to a single step in the new process [39]. In the traditional process, hydrolysis, decarboxylation, Friedel Crafts acylation, condensation, and catalytic reduction of the imine gave a mixture of cis and trans amines. Pfizer improved its manufacturing process, in which cis- and trans-sertraline were obtained by imine formation of monomethyl mine with a tetralone, followed by reduction of the imine function and in situ resolution of the diastereomeric salts of mandelic acid. The product obtained has a higher yield and a higher selectivity. In addition, a palladium catalyst was introduced in the reduction step to reduce impurities and decrease raw material consumption by 60, 40 and 20% for monomethyl amine, tetralone and mandelic acid, respectively. The implementation will save 220,000 pounds of 50% caustic soda, 330,000 pounds of 35% hydrochloric acid waste and 970,000 pounds of titanium dioxide waste annually. By eliminating waste, reducing solvents, and maximizing yields of key intermediates, Pfizer won the 2002 EPA Presidential Green Chemistry Challenge Award for Cleaner Manufacturing.

8.3 Dow AgroSciences LLC (soil pollution)

As the world population increases, so does the demand for higher crop yields, and with it the negative impact of agricultural activities on the global environment. About 75% of nitrates in the environment, for example in soil and water, come from the application of nitrogen fertilizers and manure. Basically, soil bacteria rapidly metabolize the nitrogen from the applied urea and fertilizer, resulting in less availability of nitrogen for plant nutrition. To overcome this barrier, DOW Chemical scientists are developing an aqueous microcapsule suspension product, instinct, a nitrification inhibitor to prevent nitrogen conversion from urea. Based on calculations, the use of this product reduced estimated carbon dioxide equivalent emissions by 664,000 tons and increased corn production by 50 million bushels. By eliminating toxic chemicals and CO_2, Dow Agro Chemical won the 2016 EPA Presidential Green Chemistry Challenge Award as part of a cleaner manufacturing mission.

9. Conclusion

Overall, this review has discussed the basic concept of GC and 12 principles with their application to the development of pollution free products. Along with the Principles, the associated barriers to the implementation of the GC Principles were discussed. Furthermore, this paper has also highlighted the role of PAS strategies, Prevention, Assurance and Sustainability to meet the requirements to overcome the associated barriers. In summary, the role of innovation technology and integration management would play in building GC was extensively discussed. In addition, this overview discusses the industrial case study related to the control of air, water and soil pollution by implementing the GC concept. Although the strategy discussed in this chapter is simple, it arouses enough interest among researchers to take the next step in implementing GC for a sustainable environment.

Author details

Angrui Jiang, Jingwei Li, Kinjal J. Shah* and Zhaoyang You*
College of Urban Construction, Nanjing Tech University, Nanjing, China

*Address all correspondence to: kjshah@njtech.edu.cn
and youzhaoyang@njtech.edu.cn

References

[1] Chen TL, Kim H, Pan SY, Tseng PC, Lin YP, Chiang PC. Implementation of green chemistry principles in circular economy system towards sustainable development goals: Challenges and perspectives. Science of the Total Environment. 2020;**716**(1):136998. DOI: 10.1016/j.scitotenv.2020.136998

[2] American Chemical Society National Historic Chemical Landmarks. Rachel Carson's Silent Spring. 2012. Available from: http://www.acs.org/content/acs/en/education/whatischemistry/landmarks/rachel-carson-silent-spring.html [Accessed: May 29, 2023]

[3] Luther L, Division I. CRS report for congress, The National Environmental Policy Act: Background and Implementation. 2008. Available from: https://sgp.fas.org/crs/misc/RL33152.pdf [Accessed: November 19, 2022]

[4] The Origins of EPA. 2022. Available from: https://www.epa.gov/history/origins-epa [Accessed: June 1, 2022]

[5] DDT—A Brief History and Status. 2022. Available from: https://www.epa.gov/ingredients-used-pesticide-products/ddt-brief-history-and-status [Accessed: October 18, 2022]

[6] Evaluation Report. EPA Needs a Coordinated Plan to Oversee Its Toxic Substances Control Act Responsibilities. 2010. Available from: https://www.epa.gov/sites/default/files/2015-09/documents/20100217-10-p-0066.pdf [Accessed: April 15, 2023]

[7] Green Chemistry History. 2023. Available from: https://www.acs.org/greenchemistry/what-is-green-chemistry/history-of-green-chemistry.html [Accessed: May 18, 2023]

[8] Anastas ND, Warner JC, Way S. The incorporation of hazard reduction as a chemical design criterion in green chemistry. Chemical Health & Safety. 2004;**12**(2):9-13

[9] Horváth IT, Anastas PT. Introduction: Green chemistry. Chemical Reviews. 2007;**107**(6):2167-2168

[10] Science E. Green chemistry: A framework for a sustainable future. The Journal of Organic Chemistry. 2021;**86**:8551-8555

[11] Anastas PT, Warner JC. Green Chemistry: Theory and Practice. New York: Oxford University Press; 1998

[12] Peters M, Von Der AN. It is better to prevent waste than to treat or clean up waste after it is formed – Or: What Benjamin Franklin has to do with "green chemistry". Green Chemistry. 2016;**18**:1172-1174

[13] Parent KE. Cleaning Up with Atom Economy. 2023. Available from: https://www.acs.org/content/dam/acsorg/greenchemistry/education/resources/cleaning-up-with-atom-economy.pdf [Accessed: May 18, 2023]

[14] Morsch LA, Deak L, Tiburzi D, Schuster H, Meyer B. Green aqueous Wittig reaction: Teaching green chemistry in organic teaching laboratories. Journal of Chemical Education. 2014;**91**:611-614

[15] Wittig G, Schöllkopf U. Über triphenyl-phosphin-methylene als olefinbildende Reagenzien (I. Mitteil.). Chemistry Europe. 1954;**87**(9):1318-1330

[16] Howell BA, Sun W. Biobased flame retardants from tartaric

acid and derivatives. Polymer Degradation and Stability. 2018;**157**:199-211. DOI: 10.1016/j.polymdegradstab.2018.10.006

[17] Bubalo MC, Vidovi S, Radojˇ I. Green solvents for green technologies. Journal of Chemical Technology and Biotechnology. 2015;**90**:1631-1639

[18] Kokel A, Schäfer C, Török B. Microwave-assisted reactions in green chemistry. In: Han B, Wu T, editors. Green Chemistry and Chemical Engineering. Encyclopedia of Sustainability Science and Technology Series. New York, NY: Springer; 2019. DOI: 10.1007/978-1-4939-9060-3_1008

[19] USEPS. Presidential Green Chemistry Challenge: 2012 Greener Reaction Conditions Award. 2022. Available from: https://www.epa.gov/greenchemistry/presidential-green-chemistry-challenge-2012-greener-reaction-conditions-award [Accessed: December 2, 2022]

[20] Osman AI, Fawzy S, Farghali M, El-Azazy M, Elgarahy AM, Fahim RA, et al. Biochar for agronomy, animal farming, anaerobic digestion, composting, water treatment, soil remediation, construction, energy storage, and carbon sequestration: A review. Environmental Chemistry Letters. 2022;**20**:2385-2485. DOI: 10.1007/s10311-022-01424-x

[21] Cinar SO, Chong ZK, Kucuker MA, Wieczorek N, Cengiz U, Kuchta K. Bioplastic production from microalgae: A review. International Journal of Environmental Research and Public Health. 2020;**17**(11):3842, 1-21

[22] Mehmood A, Raina N, Phakeenuya V, Wonganu B, Cheenkachorn K. The current status and market trend of polylactic acid as biopolymer: Awareness and needs for sustainable development. Materialstoday Proceedings. 2023;**72**(6):3049-3055. DOI: 10.1016/j.matpr.2022.08.387

[23] Grisales DVH, Alvarez-Aldana A, Ruales-Salcedo A, Prado-Rubio OA. 4-novel approaches toward bio-butanol production from renewable feedstocks. Advances and Development in Biobutanol Production. 2023;**5**:105-138. DOI: 10.1016/B978-0-323-91178-8.00001-1

[24] Abdulrahman RM. Green Chemistry (Principle and Practice) [Thesis]. Iraq: University of Salahaddin-Erbil; 2021. Available from: https://academics.su.edu.krd/public/profiles/suad.mohiaedin/supervision/supervision-1006-10199-1621719077-1.pdf [Accessed: November 19, 2022]

[25] Rueping M, Nachtsheim BJ. A review of new developments in the Friedel – Crafts alkylation – From green chemistry to asymmetric catalysis. Beilstein Journal of Organic Chemistry. 2010;**6**(6):1-24

[26] Shah KJ, Imae T. Photoinduced enzymatic conversion of CO2 gas to solar fuel on functional cellulose nanofiber films. Journal of Materials Chemistry A. 2017;**5**(20):9691-9701. Available from: http://xlink.rsc.org/?DOI=C7TA01861D

[27] Boethling RS, Sommer E, Difiore D. Designing small molecules for biodegradability. Chemical Reviews. 2007;**107**(6):2207-2227

[28] Shah KJ, Pan SY, Lee I, Kim H, You Z, Zheng JM, et al. Green transportation for sustainability: Review of current barriers, strategies, and innovative technologies. Journal of Cleaner Production. 2021;**326**(October):129392. DOI: 10.1016/j.jclepro.2021.129392

[29] Sjostrom J. Green chemistry in perspective — Models for GC activities

and GC policy and knowledge areas. 2006. Green Chemistry. 2006;**8**:130-137

[30] Matus KJM, Clark WC, Anastas PT, Zimmerman JB. Barriers to the implementation of green chemistry in the United States. Environmental Science and Technology. 2012;**46**:10892-10899

[31] Rana KK, Rana S. Fundamentals, representative applications and future perspectives of green chemistry: A short review. Green Chemistry. 2014;**3**(6):1-16

[32] Poliakoff M, Fitzpatrick JM, Farren TR, Anastas PT. Green chemistry: Science and politics of change. Science. 2002;**297**(5582):807-811

[33] Gupta P, Mahajan A. RSC advances green chemistry approaches as sustainable alternatives to conventional strategies in the pharmaceutical industry. RSC Advances. 2015;**5**:26686-26705

[34] Quazi S, Aleesha A, Thomas CE. A critical review on the beginning, recent advancement and upcoming challenges of green chemistry. Scholars International Journal of Chemistry and Material Sciences. 2020;**8669**:58-64

[35] Green Chemistry. A Strong Driver of Innovation, Growth, and Business Opportunity. 2021. Available from: https://greenchemistryandcommerce.org/documents/GC3GreenChemReport-ES-Nov2021.pdf [Accessed: December 2, 2022]

[36] USEPS. Presidential Green Chemistry Challenge: Award Recipient 1996-2014. 2015. Available from: https://www.epa.gov/sites/default/files/2015-02/documents/award_recipients_1996_2014.pdf [Accessed: April 15, 2023]

[37] Tirpak RA, Nabiul ARM, Winston RJ, Valenca R, Schiff K, Mohanty SK. Conventional and amended bioretention soil media for targeted pollutant treatment : A critical review to guide the state of the practice. Water Research. 2021;**189**:116648. DOI: 10.1016/j.watres.2020.116648

[38] USEPS. Presidential Green Chemistry Challenge: 2016 Designing Greener Chemicals and Specific Environmental Benefit: Climate Change Awards: Newlight Technologies. 2023. Available from: https://www.epa.gov/greenchemistry/presidential-green-chemistry-challenge-2016-designing-greener-chemicals-and-specific [Accessed: April 15, 2023]

[39] Instinct® Nitrogen Stabilizer Wins U.S. EPA Presidential Green Chemistry Challenge Award. 2023. Available from: https://www.corteva.ca/content/dam/dpagco/corteva/na/ca/en/files/guide/DF-Article-Optinyte-Global-Environoment-Guide.pdf [Accessed: April 15, 2023]

Section 2

Strategy P: Pollution and Accident Prevention

Chapter 2

Advances in the Use of Green and Sustainable Synthesis to Obtain Nanomaterials

Jessica R.P. Oliveira and Giane G. Lenzi

Abstract

The bibliometric analysis by Methodi Ordinatio reveals the impressive increase in the published articles about green chemistry, and specificity in green synthesis of nanomaterials. In the last decade, they have published over 450 articles, most led by India, China, and Iran. The green synthesis is according to the 12 principles of green chemistry (PGCs) to obtain nanoparticles with minimization of waste and toxic emissions, use of green solvents and alternatives to conventional organic solvents, use of renewable and sustainable raw materials, and energy efficiency and use of renewable energy. After synthesis, the green nanoparticles are characterized to know their physical and chemical properties. Green synthesis can contribute to the sustainable development goals (SDGs) until nine goals can be associated with green synthesis and green nanoparticle applications. Among advantages and limitations, the green syntheses of nanoparticles have the potential to grow more by future perspectives gap.

Keywords: nanomaterial synthesis, nanoparticle characterization, sustainability, green chemistry, 12 principles of green chemistry (PGC), sustainable development goals (SDG)

1. Introduction

The published increase of research articles over time about green chemistry in nanotechnology subject enriches the scientific basis with a lot of manuscripts. When the researcher needs to start any search or deepen knowledge about a determined subject, the excess of information can be a big problem. In this case, the Methodi Ordinatio (MO) can be applied to help organize the articles per relevant order [1].

Pagani *et al.* created a scientific method that considers the three main aspects of each article: impact factor, published year, and the publication's annual average of citations [1]. This method was used to classify more than 500 articles about the advances in the use of green and sustainable synthesis to obtain nanomaterials. The first hundred more relevant articles were used to write this chapter.

The most significant advances in using green and sustainable synthesis to obtain nanomaterials are not just the trend. The increased uses of microorganisms, plants, and waste in a green synthesis are related to environmental preservation. When a green synthesis of nanoparticles is performed instead of a typical synthesis, the researcher can collaborate with the 12 principles of green chemistry (PGCs) [2] and directly or indirectly to the sustainable development goals (SDGs), agenda 2030 [3].

Several bottom-up chemical or biological methods can be changed or adapted to turn a green synthesis [4]. This way, these processes can decrease waste and toxic emissions, use green and alternative solvents, use renewable and sustainable raw materials, and take advantage of energy sources better, cherishing energy efficiency and use of renewable energy [2, 5].

This chapter aims to analyze the green synthesis methods used in the first hundred articles by MO. As well as the typical characterizations needed to know about the properties of each synthesized material, situate the most application area, and establish the advantages and limitations of green synthesis of nanoparticles. The most important key of this work is compliance with the 12 principles of green chemistry in the so-called green synthesis and the close relationship between green synthesis and the objectives for sustainable development.

2. Bibliometric analysis

The importance of the bibliometric analysis is a panoramic notion about relevant publications over time. Many articles have been published and cited, but for researchers, it is complicated when starting a new work, and they need to organize these articles by importance. The MO can facilitate this first research work [1]. The MO uses one equation to calculate de InOrdinatio: the article's importance based on your impact factor, published year, and the number of citations (Eq. (1)).

$$InOrdinatio = \left\{ (\Delta * IF) - \left[\lambda * \left(\frac{ResearchYear - PubYear}{HalfLife} \right) \right] + \Omega * \frac{\sum Ci}{[(ResearchYear + 1) - PubYear]} \right\} \quad (1)$$

In the equation, the symbols represent: IF is the impact factor and corresponds to the selected journal metrics (such as JCR, CiteScore, SNIP, or SJR SCImago); Δ indicate the importance of the impact factor; λ is the relevance of the publication year; Ω is the significance about number of citations. These values, which indicate the importance of anything, can range from 1 to 10, according to the researcher's choice when using the MO in their search. The ResearchYear and PubYear are the year the research was conducted and the year the paper was published, respectively. While $\sum$Ci is the total number of citations found in Google Scholar, and halflife refers to the median cited halflife of journals with JCR 2020 [1].

In the MO, the first step is to define the keywords that will be searched and where it is done. In this case, the combination searched was ((TS = (green chemistry)) AND TS = (synthesis)) AND TS = (nanomaterials) in Web of Science, Science Direct, Scopus, and Science Open websites. These searches resulted in 6, 140, 425, and 366 articles on each site. After filtering articles to exclude duplicates that were not attractive to the work, the InOrdination equation (Eq. (1)) was applied to 509 articles. Despite many of these articles being used in writing this chapter, after the InOrdinatio ranking, the best hundred articles were chosen to base this review.

The results of this bibliometric analysis are essential to identify some data like the subject importance, publication number per year, the country where the search about the subject was happening, the gaps in the issue that need more investigation, etc. From that, insights can appear as new ideas for work, research, articles, and reviews [1].

In this context, we can note that the number of articles increases over time. The published works in the last decade are nine times smaller than the current: Have 456 published articles between 2014 and 2023 versus 50 published articles between 2007 and 2013, and this search was performed without year restriction, **Figure 1(a)**. The relevance and interest in green chemistry and nanoparticle synthesis are very significant, but India, China, and Iran are leading research in this area, **Figure 1(b)**.

The keywords used in these articles, among others, were green chemistry (37), green synthesis (24), nanoparticle (18), nanomaterial (16), photocatalysis (7), nanocomposite (6), silver nanoparticles (6), nanotechnology (6), plant (4), magnetic nanoparticles (4), plant extract (4), gold nanoparticle (4), and green (4). **Figure 2** was created using all the keywords of the hundred-first articles; the word size in the figure represents the frequency it appears in keywords.

However, green chemistry applied to nanomaterial synthesis is a current, meaningful, and exciting subject. In this sense, we need more understanding of what sustains this subject's relevance, like the principles of green chemistry, the applications of nanomaterials synthesized by green routes, and the contribution to the sustainable development goals (SDGs). These factors direct the movement's rise and research projects' political/economic interests.

3. Principles of green chemistry applied to obtaining nanomaterials

Established by Paul Anastas and John Warner in 1998, the 12 principles of green chemistry (PGCs) concern a conscious attitude guide to promoting green chemistry [2], **Figure 3**. This concept is according to ISO 14001, an international standard about the environmental management system (EMS), which applies to any organization that wishes to demonstrate conformity with this international standard by promoting sustainability and safety in your production [6].

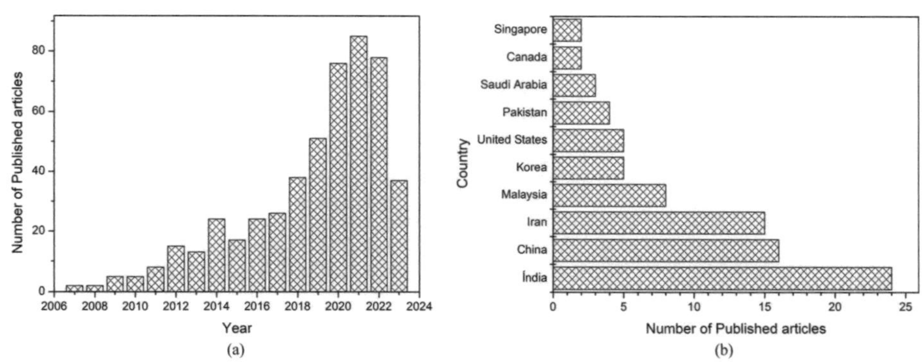

Figure 1.
Bibliometric analysis data: (a) increase in publications throughout the years, among the 509 mentioned articles, and (b) countries that published the most articles on the subject, among the 100 most important articles according to Methodi Ordinatio.

Figure 2.
Word cloud with keywords used by articles.

Currently, the "green" adjective is frequently used by researchers. This can be attributed to different causes, i.e., promoting your research, funding requirements, and creating new sustainable alternatives to a current problem or opportunity, among others [7].

Nanoscience is a concept that relates to the nanometric scale (10^{-9} m), which must be in at least one of the nanomaterials dimensions (design, manufacture, control, characterization, and applications), and the upper limit is 100 nm. The green concept is so used in nanoscience that already appears the term "Green Nano" [7].

The 12 PGCs can be associated with green nanomaterial synthesis, as the perspectives below, and all consist in performing a chemical more safely, minimizing the dangerous products or reagents, as principle 12.

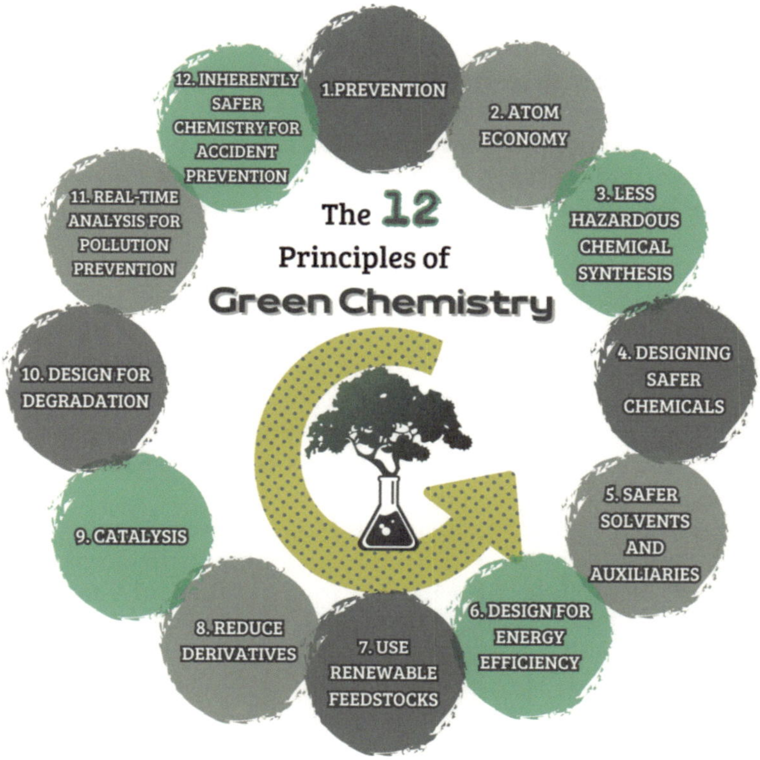

Figure 3.
Representation of the 12 principles of green chemistry proposed by P. Anastas and J. Warner.

Principle 12, "Inherently Safer Chemistry for Accident Prevention," talks about safety that is crucial within the general principles of green chemistry since risk reduction and accident prevention aim not only at environmental sustainability but also at the security of processes, workers, and communities. For this reason, PGCs 12 is contained behind all the other principles, which must be implemented by eliminating risks or at least reducing them [2, 7].

3.1 Minimization of waste and toxic emissions

The minimization of the toxicity of the process is mainly connected with PGCs 1, 2, and 11.

- Prevention: It is about reducing waste as much as possible because it can be much more difficult and expensive to solve future problems with generated waste. In synthesizing nanomaterials, it is possible to reduce the number of reagents with optimizations and avoid the generation of by-products [2, 5, 7]. Furthermore, organic residues and residues or by-products of other processes can be used as agro-wastes [8–10] and biotechnological processes [11–14].

- Atom Economy: It is not about reaction yield but about the incorporation of reagent atoms in the final product. This is a limitation in green nanomaterial synthesis. These reactions are known to cause low yield and use of several established reactants and other auxiliaries to obtain the specified nanoparticle. Despite it can be seen as an opportunity for an optimization process [2, 5, 7].

- Real-Time Analysis for Pollution Prevention: It is controlling the reaction to not produce the hazardous subproduct; it happens when trying to understand the mechanism behind the reaction. Sometimes, researchers are focused just on the final product, but they can produce any pollutants during the reaction synthesis or the washing of the nanoparticles. It is necessary to analyze the process to monitor the products of the reaction [5, 7].

3.2 Use of green solvents and alternatives to conventional organic solvents

In green nanomaterial synthesis, water-based extracts are usually used to substitute the organic solvents. It is according to three PGCs, are them:

- Less Hazardous Chemical Synthesis: This PGC means that synthesis needs to be especially careful not to harm human and planet health. In the green, nanomaterial synthesis is increasing over time, the alternative synthesis that substitutes hazardous reagents to “natural reagents” originating in plants, algae, fungi, and bacteria, among other biological entities [12, 13, 15–17].

- Designing Safer Chemicals: This PGC is complementary to the before item. Still, this one is focused on products that need safety when applied to specified functions, be it medical, chemical, industrial, or pharmaceutical, among others. The efficacy has a strict relation to security and low toxicity [5, 7].

- Safer Solvents and Auxiliaries: It is about substituting conventional organic solvents with other less dangerous/contaminants. Green nanomaterial synthesis usually can use aqueous plant extracts [18–20] and ionic liquids [21–25] for this substitution with more benefits to desired products.

3.3 Energy efficiency and use of renewable energy

The energy efficiency in green nanomaterial synthesis can be the differential about the nanomaterial synthesized; the same product can be obtained in several ways, and the economy of the energy and material can be more accessible when need expand to an industrial scale, for example.

- Design for Energy Efficiency: The economy of the energy minimizes the environmental and financial impacts. This economy can be around the reactants chosen, conditions of the temperature and pressure, and reaction way. It is common for green nanomaterial synthesis to be conducted in microwave [26–28] and ultrasound [29–32] to try to be more easily, quickly, and economized energy.

- Reduce Derivatives: This is a fragility of green nanomaterial synthesis because it is expected that derivatives (or co-products) can be created during the process [7]. It can get more waste and derivatives, especially when it needs so many steps to obtain the final product with the aim of having specific shells in its structure. The way to reduce by-products is to minimize the steps of the process. Another specific way is using enzymes [33], which are highly specific and can eliminate the use of other protecting groups or derivatives.

- Catalysis: the green nanomaterial synthesis can be catalyzed for more energy efficiency and to reduce derivatives. But we need to be careful to avoid happening the opposite. It is better that the catalyst chosen be environmentally friendly and ideally can be recovered after the reaction, and not be one more waste after the synthesis.

3.4 Use of renewable and sustainable raw materials

The use of renewable sources of raw materials and concern for the degradation of products and by-products generated are principles of green chemistry strongly related to the synthesis of new nanomaterials.

- Use of Renewable Feedstocks: As renewable raw materials, several bio-based materials have been used in nanomaterial synthesis [15, 34, 35]. The nanomaterials obtained by bio-based synthesis can be diverse, with particular physical and chemical properties.

- Design for Degradation: This can be a problem if, during the biosynthesis, a subproduct is obtained or a feedstock is hard to degrade after the reaction. So, it is necessary to design the synthesis and think about the products. The nanoparticle needs special attention because, after your use, it can be recycled, reused, recovered, or biodegraded and not be another problem to the environment [7].

4. Advances in green synthesis methods for nanomaterials

Nanomaterial synthesis can be divided into two principles: top-down and bottom-up methods [36]. In the top-down method, the bulk materials are transformed into nanoparticles by physical, chemical, and mechanical processes, which reduce the material mechanically. The main processes are spray pyrolysis, sonication, arc discharge, pulse wire discharge, pulsed laser ablation, radiation, electro-deposition, evaporation-condensation, vapor and gas phase, chemical, etching, ball milling, and lithography [13, 36–39].

The bottom-up method is the transformation of atoms/molecules into nanoparticles. This can involve the dissolution of salts in a solvent, reducing ions to their element with a reducing agent, and stabilizing the resulting neutral nanoparticles with stabilizing agents to prevent agglomerate accumulation. This process can occur by chemical methods (sol-gel technology, co-precipitation, redox processes, pyrolysis, microemulsion, microwave, photochemical, electrochemical, sonochemical, and hydrothermal activity) and biological methods (using fungi, plants, yeast, bacteria, viruses, and biomolecules/biopolymers) [13, 36–39].

Several methods can be adapted to be "green" by applying the PGCs. Currently, biological base materials are used in the synthesis process, like plants, biomass, microorganisms (bacteria, fungi, and yeast), and enzymes, among others, like additives, stabilizers, or substitutes for other harmful compounds. The aim is the same: obtain the better synthesis method, cheapest, quickly executed, and most sustainable.

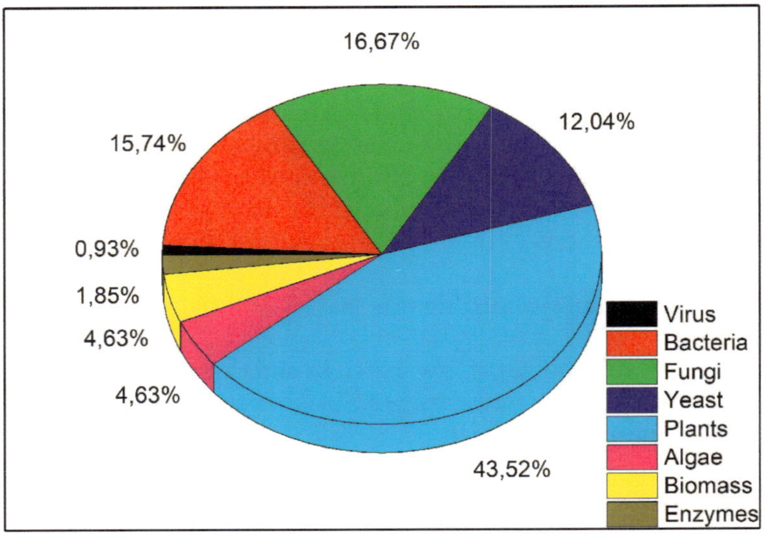

Figure 4.
Representativeness of the types of raw materials used in the first hundred articles with the best InOrdinatio.

Among the principal articles according to InOrdinatio classification, the plants are the feedstocks more used in nanomaterial synthesis, allowed by fungi, bacteria, and yeast (**Figure 4**). It can be explained by the diversity of the plants and part of it (leaf, stem, flower, fruit, peel, and seed) that can be used in synthesis.

4.1 Nanomaterial synthesis using plant extracts

The increase in the use of plant extracts is explained by possible variations with different plants and parts of it. This variation is related to the influence caused by the secondary metabolites, a particular characteristic of each species [38, 40]. Usually, different metallic nanoparticles with different sizes can be synthesized using plant extracts, **Table 1**, in which cadmium, chromium, copper, gold, iron, silver, and zinc nanoparticles are the most common.

Another typical process is using biomass from any part of the plant in nanomaterial synthesis to obtain the nanoparticles, like the seed, fruit peel, fallen leaves, and bagasse.

4.2 Use of microorganisms for the nanomaterial synthesis

It is increasing the use of microorganisms in nanomaterial synthesis [12, 15, 34, 61, 62]. Bacteria, fungi, and yeast, when added have more expressive results than plants in nanomaterials synthesis (**Figure 5**). The bacteria application is related to the bacteria resistance with variations in temperature, pH, incubation time, and oxygenation, among other environmental stresses [15, 34, 61]. Another cause of the success of the bacteria is the handling, manipulation, and genetic modification facilitated [61]. Usually, these adjustments depend on the size, shape, and composition of interest [15]. Usually, metallic nanoparticles are synthesized with bacteria (Au, Ag, CdS, Co_3O_4, Fe_3O_2, Hg, MnO_2, PbS, Pd, Pt, Sb, Se, Te, Ti, ZnS, and ZrO_2 nanoparticles) [13, 34, 61, 62], which can be synthesized by either intracellular or extracellular mechanisms [12, 15].

Nanomaterial obtained	Size (nm)	Plant used	Reference
Cadmium oxide	113	*Hibiscus Sabdariffa* flower extract	[41]
Chromium Oxide	17–42	*Abutilon indicum (L.)* leaf extract	[42]
Copper nanoparticles	5–20	*Ziziphus spina-christi (L.)* fruit extract	[43]
Copper oxide	61.48	*Momordica charantia* fruit extract	[44]
Gold Silver	15–25 15–20	*Memecylon umbellatum* leaf extract	[45]
Gold Silver	5.82 5.54	*Clerodendrum inerme* leaves extract	[46]
Gold nanoparticles	10–100	*Nerium oleander* stem bark extract	[47]
Gold nanoparticles	5–20	*Gnidia glauca* flower extract	[48]
Gold nanoparticles	20–140	Rind of watermelon aqueous extract	[49]
Iron oxide	13.42	*Garcinia mangostana* fruit peel extract	[50]
Iron oxide	20–30	*Excoecaria cochinchinensis* extract	[51]
Silver nanoparticles	18.2	*Vitex negundo L.* extract	[52]
Silver nanoparticles	37.71–71.99	*Chrysanthemum indicum L.*	[53]
Silver nanoparticles	21–173	*Conocarpus Lancifolius* fruits extract	[54]
Silver nanoparticles	25.2	Nigella sativa extract	[55]
Silver nanoparticles	45–110	*Brillantaisia patula* extract *Crossopteryx febrífuga* extract *Senna siamea* extract	[56]
Silver oxide Zinc oxide	50 60	*Moringa oleifera* gum	[57]
Zinc Oxide	52.24	*Cayratia pedata* leaves extract	[58]
Zinc oxide	480	*Panax ginseng* extract *Acanthopanax senticosus* extract Kalopanax septemlobus extract *Dendropanax morbifera* extract	[59]
Zinc oxide	9–38	*Azadirachta indica* extract	[60]

Table 1.
Metallic nanoparticles synthesized with different plant extracts.

The nanomaterial synthesis using fungi has interesting advantages because of the presence of enzymes, proteins, and/or reducing components on their cell surfaces [13, 15]. Fungi have unique characteristics such as their resistance to toxicity, ease of handling, large surface areas that lead to increased production rates, easy and simple downstream processing, economic viability, and the larger spectrum of nanoparticle morphologies [12, 34, 61]. Furthermore, the synthesis process can be facilitated because they endure harsh synthesis conditions such as flow pressure or agitation in the bioreactor, and have accelerated growth in controllable ways, and almost all fungi are better resistant to genetic or environmental mutations when compared with bacteria [12, 61]. Both extracellular and intracellular syntheses of metallic nanoparticles

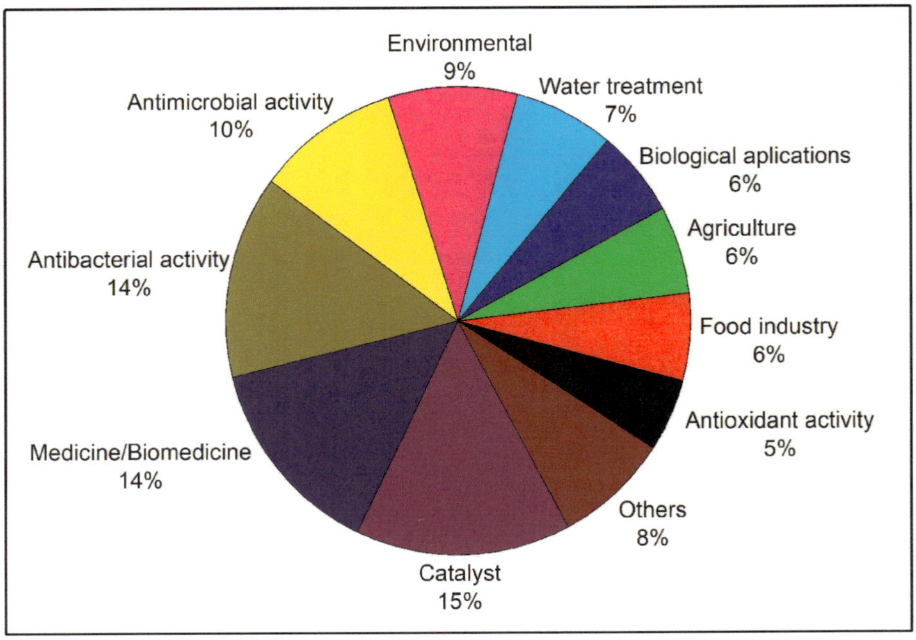

Figure 5.
Where the nanomaterials obtained by green synthesis are applied in the first hundred articles with greater relevance to this research.

using fungi have been investigated as Au, Ag, TiO_2, ZnO, and ZrO_2 nanoparticles [13, 34, 62].

Yeasts, which are eukaryotic microorganisms, like other microorganisms, have been explored in nanomaterial synthesis [13, 15, 34]. Their ability to absorb and accumulate toxic metals enables them to be used in the synthesis of nanomaterials [15, 34]. Diverse yeast species are employed for the preparation of innumerable metallic nanoparticles, such as Ag, Au, CdS, Fe_3O_4, PbS, and Sb_2O_3 [13, 34]. During the synthesis, the mechanism used by yeast to form and stabilize the nanoparticle defines its size, shape, and properties [15].

4.3 Application of green methods in nanoparticle synthesis

Several eco-friendly techniques can be used for nanomaterial synthesis: sonochemical, microwave, hydrothermal, solvothermal, and electrochemical, among others. But two of them are the most used: sonochemical [29, 36, 63, 64] and microwave [26–28, 36, 63] methods.

The sonochemical method uses ultrasound, an inaudible sound wave with a frequency higher than 20 kHz [63]. The ultrasound possesses high energy that can generate the acoustic cavitation effect, which results consecutively in the formation, growth, and implosive collapse of these bubbles in a liquid environment. During the collapse of a cavity, high local temperatures and pressures arise within a very short period, which leads to an increase in the rate of reactions [29, 36, 63, 64]. It is an eco-friendly, green, fast, and easy method of nanostructure synthesis. It is used for nanostructure material preparation and unusual nanostructured inorganic materials such as carbonyl compounds ($Fe(CO)_5$, $Co(CO)_3NO$, $Mo(CO)_6$, and $W(CO)_6$) [36].

In microwave-assisted synthesis, electromagnetic irradiation can reduce the energy needed for the synthesis process, providing rapid heating and facilitating greener preparation of nanoparticles [26, 27]. The main advantages of using microwave-assisted in nanomaterial synthesis are reduced reaction time (causes the reduction in activation energy), high product yield (a shorter reaction time reduces the chance of undesirable side products), high product purity, high reproducibility (a uniform microwave field around the reaction mixture), and reaction conditions can be easily optimized [26]. This technique is used to synthesize various metal nanoparticles [36].

5. Characterization of nanomaterials obtained by green synthesis

The nanomaterials need characterizations to confirm your structure and chemical and physical properties. Among them are the most common to observe about the structure, such as transmission electron microscopy (TEM), X-ray diffraction (XRD), and scanning electron microscopy (SEM). Other characterizations can be needed to evaluate specific chemical and physical properties according to the components and applications of the synthesized nanomaterials, such as tests to assess the stability and reactivity.

5.1 Structural characterization techniques

5.1.1 Transmission electron microscopy

The transmission electron microscopy (TEM) characterization is the most popular among the nanoparticles [65–67]. With this analysis, the size and shape of the nanomaterials can be evaluated. And, if it has a coating or void, it can be observed.

The TEM is a powerful electron microscope (the magnification can be 2 million times better than that of the light microscope) and uses a beam of electrons to focus on a sample, producing a highly detailed image with morphological features, compositions, and crystallization information. The principle of TEM is the beam of electrons, which has a wavelength shorter than light. When the electron illuminates the sample, the resolution power increases, consequently increasing the wavelength of the electron transmission [68].

TEM analysis has applications in nanotechnology to study nanoparticles and detect and identify fractures and damaged microparticles, enabling subsequent repair mechanisms for these particles. Among other fields, TEM can be applied in biology, microbiology, and forensic studies to examine the cell structures of plants, animals, and microorganisms and analyze bacteria flagella, plasmids, and the shapes/sizes of microbial cell organelles [68].

The TEM have limitations, such as the price, the extensive equipment, the tedious sample preparation, and laborious operation and maintenance because of its sensibility. But, despite the limitations, it has important advantages to researchers, like significant and power magnification, the variety of fields that can be applied, and the production of very efficient, high-quality images with high clarity [68].

5.1.2 X-ray diffraction

X-ray diffraction (XRD) is a quick and nondestructive analytical technique that can be applied to powder, solid, and liquid samples to analyze their physical properties. With this characterization it is possible to know about the composition, cell dimensions, and crystalline structure [69, 70].

The principle of XRD is based on constructive interference between monochromatic X-ray and a crystalline sample. The X-ray is generated by a cathode ray tube. These X-rays are filtered, collimated, and directed to the sample. The X-rays' wavelength is the same magnitude as the distance between the atoms in a crystalline lattice [69, 70]. The constructive interference and a diffracted ray produce a diffraction pattern, which satisfies Bragg's law ($n\lambda = 2d\sin\theta$), where n is a natural number, λ is the wavelength of the incident radiation, d is the distance between atomic planes, and θ is the angle of incidence about the considered plane. This law relates the wavelength of electromagnetic radiation to the diffraction angle and the lattice spacing in a crystalline sample, and it is used in the measurement of crystals and their phases [69, 70].

The XRD is essential to develop and improve new materials in laboratories and industries. The main uses are qualitative and quantitative phase analysis of pure substances and mixtures. The most common method for phase analysis is often called X-ray powder diffraction (XRPD). The XRPD is used to characterize crystalline materials, identify fine-grained minerals that are difficult to determine optically, determine unit cell dimensions, and measure sample purity. With enhancements, it is possible to determine crystal structures using Rietveld refinement, determine modal amounts of minerals (quantitative analysis), and characterize thin film samples, among others [69, 70].

The XRD advantages are that, in most cases, it provides an unambiguous determination, sample preparation is minimal, and the data interpretation is relatively straightforward (the operator does not need to be an XRD expert), which makes this technique the most popular and suitable [69, 70]. Usually, the researchers use the standards published by the Joint Committee on Powder Diffraction Standards-International Center for Diffraction Data (JCPDS-ICDD) to help identify the composition of their sample.

5.1.3 Scanning electron microscopy

The scanning electron microscope (SEM) is a type of electron microscope that produces images that possibility study solid particles surfaces. With the increased knowledge about nanotechnology, the materials continue to shrink, making this analysis increasingly necessary for their characterization [71, 72]. The SEM principles consist of using electrons for imaging in a similar way that light microscopes use visible light, but the resolution of SEM is superior to that of a light microscope because the wavelength of electrons is much smaller than that of light [71].

The SEM uses a focused beam of electrons to generate a variety of signals that scan the surface of a solid sample. The signals that derive from electron-sample interactions reveal sample information, such as external morphology, chemical composition, and crystalline structure and orientation of materials making up the sample [71–73].

Among the advantages of SEM are that no elaborate sample preparation techniques are required and the capability to perform analyses of specific points on the sample [72, 73]. With this technique, it is possible to generate high-resolution images of the shapes of objects and to show spatial variations in chemical compositions by obtaining elemental maps or spot chemical analyses using energy-dispersive spectroscopy (EDS), discrimination of phases based on the mean atomic number using backscattered electrons (BSE), and compositional maps based on differences in trace element activators using cathode-luminescence (CL) [73].

6. Applications of nanomaterials obtained by green synthesis

The nanoparticles synthesized by green routes can have a lot of applications. Among the hundred articles with the higher InOrdinatio, the areas that aroused interest in nanotechnology applications were catalysis [28, 29, 48, 59, 60, 64, 74–80], medical/biomedical [19, 42, 45, 49, 50, 62, 63, 81–86], environmental [7, 27, 34, 63, 87–89], biological [33, 46, 56, 90, 91], water treatment [27, 37, 38, 89, 92–94], agriculture [18, 34, 95–97], and food industries [37, 38, 87, 95, 97], among others [98–103]. Several researches do not establish a specific application but show the characteristics of the nanoparticles, like antimicrobial [61, 104] or antibacterial [43, 46, 49, 53, 57, 60, 105, 106] activity, as a differential between these materials (**Figure 5**).

The efficiency in a determined application depends on the composition, physical and chemical properties, and sometimes on the particular characteristics of the nanoparticles. It is common for the same nanoparticle to be applied with different functions in different areas, with some needed adjustments. Most can be versatile, while others are specific to be efficient in just one application.

7. Contribution to the sustainable development goals (SDGs)

In 2015, the United Nations adopted the sustainable development goals (SDGs) with social, environmental, economic, and institutional dimensions. The SDGs aim to end poverty, protect the planet, and promote peace and prosperity by 2030. The United Nations General Assembly (UNGA) has set 17 integrated goals, which the results must affect and balance social, economic, and environmental sustainability. The SDGs must be achieved in every context [3].

In green chemistry, namely, the synthesis of nanomaterials can involve several of the 17 SDGs, such as 3. good health and well-being; 6. clean water and sanitation; 7. affordable and clean energy; 8. decent work and economic growth; 9. industry, innovation, and infrastructure; 11. sustainable cities and communities; 12. responsible consumption and production; 13. climate action; and 14. life below water (**Figure 6**).

Each objective mentioned will be related below to the goals that the green synthesis of nanomaterials can collaborate directly or indirectly.

- Good health and well-being: The nanoparticle synthesis is aligned with some goal targets, such as: "substantially reduce the number of deaths and illnesses from hazardous chemicals and air, water and soil pollution and contamination", and "Support the research and development of vaccines and medicines for the communicable and non-communicable diseases that primarily affect developing countries" [3]. Several nanoparticles synthesized by green routes used environmentally friendly feedstock rather than hazardous and contaminant products. Furthermore, many nanoparticles can be applied in medicine or nanomedicine, like a disease treatment.

- Clean water and sanitation: Considering that 80% of the used water does not receive adequate treatment, the majority of nanoparticles synthesized by green chemistry can be applied in water treatment or catalysis to degrade nocive compounds. So, the nanomaterial is closely related to this goal, specifically with the target "improve water quality by reducing pollution, eliminating dumping

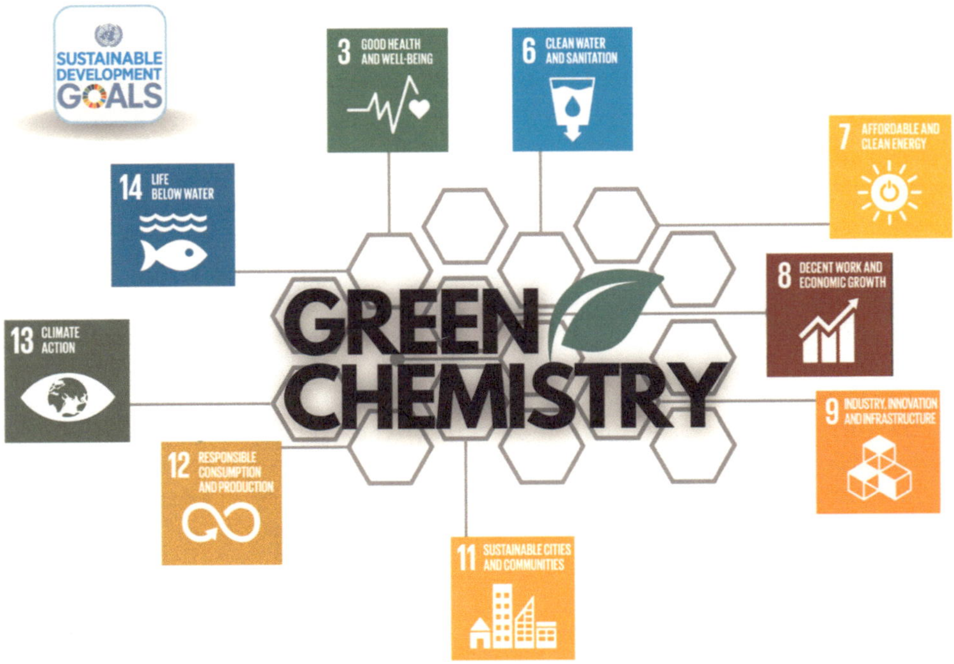

Figure 6.
SDGs related to green chemistry in the synthesis of nanomaterials.

and minimizing release of hazardous chemicals and materials, halving the proportion of untreated wastewater and substantially increasing recycling and safe reuse globally" [3].

- Affordable and clean energy: When talking about clean energy involved in nanomaterials synthesis, we can talk about the energy used in nanoparticle synthesis or nanomaterials that enable or facilitate the use of renewable energies. In this context, it can establish a relation with the target "to facilitate access to clean energy research and technology, including renewable energy, energy efficiency and advanced and cleaner fossil-fuel technology, and promote investment in energy infrastructure and clean energy technology" [3].

- Decent work and economic growth: The research investment in nanotechnology (or any research area) collaborates by itself with decent work and economic growth because it can "Promote development-oriented policies that support productive activities, decent job creation, entrepreneurship, creativity and innovation" and "reduce the proportion of young people not in employment, education or training" [3]. The green nanoparticles can be associated with goal targets: "endeavor to decouple economic growth from environmental degradation" and "promote safe and secure working environments for all workers" [3].

- Industry, innovation, and infrastructure: With the enhancement of the research is possible to meet the targets: "Support domestic technology development, research and innovation in developing countries" and "Enhance scientific research, upgrade the technological capabilities of industrial sectors in all countries, in particular developing countries" [3]. With green nanotechnology is possible to collaborate to "upgrade infrastructure and retrofit industries to make them sustainable, with increased resource-use efficiency and greater adoption of clean and environmentally sound technologies and industrial processes, with all countries taking action in accordance with their respective capabilities" [3].

- Sustainable cities and communities: Synthesizing nanoparticles by a green process can contribute to targets "Strengthen efforts to protect and safeguard the world's natural heritage" and "reduce the adverse per capita environmental impact of cities" [3]. It is because the green synthesis aims for the 12 PGCs, to trying not to pollute before, during, and after the process, besides using some green nanoparticles in environmentally friendly applications.

- Responsible consumption and production: The use the environmental-friendly feedstock in green synthesis of nanoparticles, such as food waste, contributes to goal 12 in specific targets such as "achieve the sustainable management and efficient use of natural resources," "reduce food losses along production and supply chains, including postharvest losses," "achieve the environmentally sound management of chemicals and all wastes throughout their life cycle," and "substantially reduce waste generation through prevention, reduction, recycling and reuse" [3]. When a researcher changes one chemical in your synthesis by a waste, your synthesis can be more sustainable and further one or more targets cited.

- Climate action: Green synthesis favors the environment's health, indirectly affecting climate health. When promoting research based on green synthesis of nanomaterials can promote the target "Improve education, awareness-raising and human and institutional capacity on climate change mitigation, adaptation, impact reduction, and early warning" [3].

- Life below water: Over time, the use of nanotechnology in water treatment is increasing. Consequently, in better water, the capacity of the life quality is better. By green synthesis, the quality of the water can be preserved without discard of hazardous chemicals, and the water quality can be recovered by water treatment using green nanotechnology. This way the targets achieved can be "prevent and significantly reduce marine pollution of all kinds, in particular from land-based activities, including marine debris and nutrient pollution" and "Minimize and address the impacts of ocean acidification, including through enhanced scientific cooperation at all levels" [3].

This way, green chemistry should not be seen just as a trend but as an alternative solution for the search for a more sustainable world.

8. Advantages, limitations, and future perspectives of the nanomaterial obtained by green synthesis

8.1 Advantages and limitations

The nanomaterials obtained by green synthesis have several advantages, such as no use of hazardous chemicals, reuse of different wastes, possibility to application in several areas, high specificity, less toxic effect, the feasibility of the approach, eco-friendly, cost-effective, among others [27, 38, 103].

When are using plant extracts the nanoparticles can be synthesized and functionalized by the same compound, decreasing one or more steps to it. The micro-organisms can favor the resulting selectivity [27]. Several nanomaterials have excellent antioxidant and antimicrobial action besides body biocompatibility, which can be used in medicine and biomedicine applications. Green synthesis can collaborate with this particular characteristic [27, 38, 103].

Basically, the potential of green nanotechnology includes the "natural" feedstock used, nonuse and nonelimination of hazardous chemicals, easy process that possibility reproducibility, application of the principles of green chemistry, and care with the eco and human health [28, 34].

However, the several advantages of green-synthesized nanomaterials do not exclude their limitation, which can be the costs associated with manufacturing and processing on a large scale, the unknown toxicity, improper management after use, and the unknown reaction with other compounds after discard [34, 38, 93]. The incorrect manipulation of the nanomaterials, independent of their synthesis method, can cause an environmental risk instead of a solution and directly and indirectly affect soil, fauna, and humans [93].

Applying the nanoparticles with security requires several characterizations about their physical and chemical properties and surface and their behavior in different conditions, thus having a high cost. Another limitation is the unregulated application of the nanoparticles, which can pose potential risks to human and environmental health, mainly when applied in agrifood, cosmetic, and medicine industries [34, 93, 107].

The limitations need to be overcome so the nanoparticles collaborate with both principles of green chemistry and sustainable development goals. Otherwise, the synthesis can be green by PGC and SDG, but the materials obtained will go against hand.

8.2 Future perspectives

There has been an undiscussed expressive increase in the use of green synthesis for nanoparticles in the last few years. The more than 456 articles published in the last decade show it. However, there are gaps when discussing green synthesis, which needs more studies, exploration, and scientific knowledge. Future perspectives can collaborate to optimize and use synthesis and nanoparticles obtained, such as [37, 39]:

1. Many nonexplored natural materials can be used in green synthesis (biopolymers, plants, microorganisms, among others), mainly the nearby/regional feedstocks.

2. It needs better control of the size and shape of nanoparticles. Several synthesis methods possibilities the average size between an interval but not a standard size. Some applications require more precision.

3. It is very important know-how the behavior during the synthesis and about nanoparticles. The possible mechanisms can happen and secondary mechanisms if

they exist. This makes synthesis methods predictable and consistent, even for producing the same material. Understanding the biological components, chemical agents, and molecular mechanisms involved in synthesis is also required.

4. The green synthesis is still primarily restricted to the laboratory phase. Scientific investigations are required to implement the industrial production of green nanomaterials.

5. Effectiveness risk management is necessary for the whole life cycle of green synthesis (production, handling, storage, and disposal). The toxic nature of nanomaterials is a serious concern.

9. Conclusion

In conclusion, the remarkable increase in published papers on green chemistry, particularly in the context of sustainable synthesis of nanomaterials, highlights the global commitment to more ecologically correct scientific practices. This trend emphasizes a concerted quest to adopt green principles for synthesizing nanomaterials.

By adhering to the 12 principles of green chemistry, researchers have explored the potential of green synthesis techniques, marking a new era in nanoparticle manufacturing characterized by minimal waste generation, reduced toxic emissions, and a shift toward renewable resources and sustainable processes. Integrating green synthesis practices for nanomaterials aligns with the United Nations Sustainable Development Goals (SDGs), significantly contributing to at least nine fundamental goals.

Despite the advantages, it is vital to recognize the limitations and challenges accompanying nanoparticle green synthesis. However, these limitations should clarify this field's potential for growth and innovation. After all, the scientific community continues to explore, refine, and expand the boundaries of green synthesis. The future of the synthesis of green nanomaterials is promising, inviting us to explore its potential more profoundly and contribute significantly to a more sustainable world.

Acknowledgements

This study was financed in part by the Coordenação de Aperfeiçoamento de Pessoal de Nível Superior – Brasil (CAPES) – Finance Code 001.

Conflict of interest

The authors declare no conflict of interest.

Author details

Jessica R.P. Oliveira* and Giane G. Lenzi
Department of Post-graduation in Production Engineering, Federal University of Technology – Paraná (UTFPR) Campus Ponta Grossa, Rua Doutor Washington Subtil Chueire, CEP, Ponta Grossa, Brazil

*Address all correspondence to: jeh.rocio@gmail.com

References

[1] Pagani RN, Pedroso B, dos Santos CB, Picinin CT, Kovaleski JL. Methodi Ordinatio 2.0: Revisited under statistical estimation, and presenting finder and rank in. Quality and Quantity. 2022;**57**: 4563-4602. DOI: 10.1007/s11135-022-01562-y

[2] ACS Chemistry for life, 12 Principles of Green Chemistry. [Internet]. Available from: https://www.acs.org/greenchemistry/principles/12-principles-of-green-chemistry.html [Accessed: July 8, 2023]

[3] United Nations Development Programme, THE SDGS IN ACTION [Internet]. 2015. Available from: https://www.undp.org/sustainable-development-goals?gclid=Cj0KCQjwnrmlBhDHARIsADJ5b_njdNkJH46XlfGzeft_s8lwrll8WsqfeBfcvkkX4xuHFeYD0LKDz64aAhUiEALw_wcB [Accessed: July 12, 2023]

[4] Singh J, Kaur H, Rawat M. A novel green approach for the synthesis of tungsten oxide nanorods and its efficient potential towards photocatalytic degradation of reactive green 19 dye. Journal of Materials Science: Materials in Electronics. 2018;**29**: 13715-13722. DOI: 10.1007/s10854-018-9501-6

[5] Compound interest, the twelve principles of green chemistry: What it is, & Why it Matters. 2015. [Internet]. https://www.compoundchem.com/2015/09/24/green-chemistry/ [Accessed: July 10, 2023]

[6] ISO, International Standard - ISO 14001, Switzerland. 1996. [Internet]. Available from: https://corpslakes.erdc.dren.mil/employees/envcomp/pdfs/iso-14001.pdf

[7] García-Quintero A, Palencia M. A critical analysis of environmental sustainability metrics applied to green synthesis of nanomaterials and the assessment of environmental risks associated with the nanotechnology. The Science of the Total Environment. 2021; **793**:148524. DOI: 10.1016/j.scitotenv.2021.148524

[8] Testa ML. Functionalized nanomaterials for biomass conversion. Materials Today Proceedings. 2021;**35**: 156-163. DOI: 10.1016/j.matpr.2020.04.064

[9] Kirti S, Bhandari VM, Jena J, Bhattacharyya AS. Elucidating efficacy of biomass derived nanocomposites in water and wastewater treatment. Journal of Environmental Management. 2018; **226**:95-105. DOI: 10.1016/j.jenvman.2018.08.028

[10] Omran BA, Baek K-H. Valorization of agro-industrial biowaste to green nanomaterials for wastewater treatment: Approaching green chemistry and circular economy principles. Journal of Environmental Management. 2022;**311**: 114806. DOI: 10.1016/j.jenvman.2022.114806

[11] Amina M, Al Musayeib NM, Alarfaj NA, El-Tohamy MF, Oraby HF, Al Hamoud GA, et al. Biogenic green synthesis of MgO nanoparticles using Saussurea costus biomasses for a comprehensive detection of their antimicrobial, cytotoxicity against MCF-7 breast cancer cells and photocatalysis potentials. PLoS One. 2020;**15**:e0237567. DOI: 10.1371/journal.pone.0237567

[12] Narayanan KB, Sakthivel N. Biological synthesis of metal nanoparticles by microbes. Advances in

Colloid and Interface Science. 2010;**156**: 1-13. DOI: 10.1016/j.cis.2010.02.001

[13] Singh J, Dutta T, Kim K-H, Rawat M, Samddar P, Kumar P. 'Green' synthesis of metals and their oxide nanoparticles: Applications for environmental remediation. Journal of Nanobiotechnology. 2018;**16**:84. DOI: 10.1186/s12951-018-0408-4

[14] Hasan M, Ullah I, Zulfiqar H, Naeem K, Iqbal A, Gul H, et al. Biological entities as chemical reactors for synthesis of nanomaterials: Progress, challenges and future perspective. Materials Today Chemistry. 2018;**8**:13-28. DOI: 10.1016/j.mtchem.2018.02.003

[15] Shah M, Fawcett D, Sharma S, Tripathy SK, Poinern GEJ. Green synthesis of metallic nanoparticles via biological entities. Materials (Basel). 2015;**8**:7278-7308. DOI: 10.3390/ma8115377

[16] Ahmad S, Munir S, Zeb N, Ullah A, Khan B, Ali J. Green nanotechnology: A review on green synthesis of silver nanoparticles — An ecofriendly approach. International Journal of Nanomedicine. 2019;**14**:5087-5107. DOI: 10.2147/IJN.S200254

[17] Singh A, Gautam PK, Verma A, Singh V, Shivapriya PM, Shivalkar S, et al. Green synthesis of metallic nanoparticles as effective alternatives to treat antibiotics resistant bacterial infections: A review. Biotechnology Reports. 2020;**25**:e00427. DOI: 10.1016/j.btre.2020.e00427

[18] Castillo-Henríquez L, Alfaro-Aguilar K, Ugalde-Álvarez J, Vega-Fernández L, de Oca-Vásquez GM, Vega-Baudrit JR, et al. Green synthesis of gold and silver nanoparticles from plant extracts and their possible applications as antimicrobial agents in the agricultural area. Nanomaterials. 2020;**10**:1763. DOI: 10.3390/nano10091763

[19] Mousavi SM, Hashemi SA, Ghasemi Y, Atapour A, Amani AM, Savar Dashtaki A, et al. Green synthesis of silver nanoparticles toward bio and medical applications: Review study. Artificial Cells, Nanomedicine, and Biotechnology. 2018;**46**:S855-S872. DOI: 10.1080/21691401.2018.1517769

[20] Kumar H, Bhardwaj K, Kuča K, Kalia A, Nepovimova E, Verma R, et al. Flower-based green synthesis of metallic nanoparticles: Applications beyond fragrance. Nanomaterials. 2020;**10**:766. DOI: 10.3390/nano10040766

[21] Saquib M, Ahamad S, Khan MF, Khan MI, Hussain MK. An ultrasound assisted, ionic liquid-molecular iodine synergy driven efficient green synthesis of pyrrolobenzodiazepine-triazole hybrids as potential anticancer agents. Frontiers in Pharmacology. 2023;**14**:01-11. DOI: 10.3389/fphar.2023.1168566

[22] Pandiyan N, Murugesan B, Arumugam M, Sonamuthu J, Samayanan S, Mahalingam S. Ionic liquid - a greener templating agent with Justicia adhatoda plant extract assisted green synthesis of morphologically improved Ag-Au/ZnO nanostructure and it's antibacterial and anticancer activities. Journal of Photochemistry and Photobiology B: Biology. 2019;**198**: 111559. DOI: 10.1016/j.jphotobiol.2019.111559

[23] Sameri F, Bodaghifard MA, Mobinikhaledi A. Ionic liquid-coated nanoparticles (CaO@SiO2@BAIL): A Bi-functional and environmentally benign catalyst for green synthesis of pyridine, pyrimidine, and Pyrazoline derivatives. Polycyclic Aromatic Compounds. 2022;**42**:4700-4716. DOI: 10.1080/10406638.2021.1903954

[24] Norouzi M, Elhamifar D. Ionic liquid-modified magnetic mesoporous silica supported tungstate: A powerful and magnetically recoverable nanocatalyst. Composites. Part B, Engineering. 2019;**176**:107308. DOI: 10.1016/j.compositesb.2019.107308

[25] Avirdi E, Hooshmand SE, Sepahvand H, Vishwanathan V, Bahadur I, Katata-Seru LM, et al. Ionic liquids-assisted greener preparation of silver nanoparticles. Current Opinion in Green and Sustainable Chemistry. 2022;**33**:100581. DOI: 10.1016/j.cogsc.2021.100581

[26] Kumar A, Kuang Y, Liang Z, Sun X. Microwave chemistry, recent advancements, and eco-friendly microwave-assisted synthesis of nanoarchitectures and their applications: A review. Materials Today Nano. 2020;**11**:100076. DOI: 10.1016/j.mtnano.2020.100076

[27] Mondal P, Anweshan A, Purkait MK. Green synthesis and environmental application of iron-based nanomaterials and nanocomposite: A review. Chemosphere. 2020;**259**:127509. DOI: 10.1016/j.chemosphere.2020.127509

[28] Varma RS. Greener approach to nanomaterials and their sustainable applications. Current Opinion in Chemical Engineering. 2012;**1**:123-128. DOI: 10.1016/j.coche.2011.12.002

[29] Safari J, Javadian L. Ultrasound assisted the green synthesis of 2-amino-4H-chromene derivatives catalyzed by Fe3O4-functionalized nanoparticles with chitosan as a novel and reusable magnetic catalyst. Ultrasonics Sonochemistry. 2015;**22**:341-348. DOI: 10.1016/j.ultsonch.2014.02.002

[30] Zhang Y, Zhang N, Wang Q, Wang P, Yu Y. A facile and eco-friendly approach for preparation of microkeratin and nanokeratin by ultrasound-assisted enzymatic hydrolysis. Ultrasonics Sonochemistry. 2020;**68**:105201. DOI: 10.1016/j.ultsonch.2020.105201

[31] García DA, Mendoza L, Vizuete K, Debut A, Arias MT, Gavilanes A, et al. Sugar-mediated green synthesis of silver selenide semiconductor nanocrystals under ultrasound irradiation. Molecules. 2020;**25**:5193. DOI: 10.3390/molecules25215193

[32] Gao L, Mei S, Ma H, Chen X. Ultrasound-assisted green synthesis of gold nanoparticles using citrus peel extract and their enhanced anti-inflammatory activity. Ultrasonics Sonochemistry. 2022;**83**:105940. DOI: 10.1016/j.ultsonch.2022.105940

[33] Fahimeh Charbgoo MBA, Darroudi M, Charbgoo F, Ahmad M, Darroudi M, Fahimeh Charbgoo MBA, et al. Cerium oxide nanoparticles: Green synthesis and biological applications. International Journal of Nanomedicine. 2017;**12**:1401-1413. DOI: 10.2147/IJN.S124855

[34] Saratale RG, Karuppusamy I, Saratale GD, Pugazhendhi A, Kumar G, Park Y, et al. A comprehensive review on green nanomaterials using biological systems: Recent perception and their future applications. Colloids Surfaces B Biointerfaces. 2018;**170**:20-35. DOI: 10.1016/j.colsurfb.2018.05.045

[35] Nasaruddin RR, Chen T, Yao Q, Zang S, Xie J. Toward greener synthesis of gold nanomaterials: From biological to biomimetic synthesis. Coordination Chemistry Reviews. 2021;**426**:213540. DOI: 10.1016/j.ccr.2020.213540

[36] Jamkhande PG, Ghule NW, Bamer AH, Kalaskar MG. Metal

nanoparticles synthesis: An overview on methods of preparation, advantages and disadvantages, and applications. Journal of Drug Delivery Science and Technology. 2019;**53**:101174. DOI: 10.1016/j.jddst.2019.101174

[37] Ahmed SF, Mofijur M, Rafa N, Chowdhury AT, Chowdhury S, Nahrin M. Green approaches in synthesising nanomaterials for environmental nanobioremediation: Technological advancements, applications, benefits and challenges. Environmental Research. 2022;**204**:111967. DOI: 10.1016/j.envres.2021.11196

[38] Mustapha T, Misni N, Ithnin NR, Daskum AM, Unyah NZ. A review on plants and microorganisms mediated synthesis of silver nanoparticles, role of plants metabolites and applications. International Journal of Environmental Research and Public Health. 2022;**19**:674. DOI: 10.3390/ijerph19020674

[39] Kumar JA, Krithiga T, Manigandan S, Sathish S, Renita AA, Prakash P, et al. A focus to green synthesis of metal/metal based oxide nanoparticles: Various mechanisms and applications towards ecological approach. Journal of Cleaner Production. 2021;**324**:129198. DOI: 10.1016/j.jclepro.2021.129198

[40] Marslin G, Siram K, Maqbool Q, Selvakesavan R, Kruszka D, Kachlicki P, et al. Secondary metabolites in the green synthesis of metallic nanoparticles. Materials (Basel). 2018;**11**:940. DOI: 10.3390/ma11060940

[41] Thovhogi N, Park E, Manikandan E, Maaza M, Gurib-Fakim A. Physical properties of CdO nanoparticles synthesized by green chemistry via hibiscus sabdariffa flower extract. Journal of Alloys and Compounds. 2016;**655**:314-320. DOI: 10.1016/j.jallcom.2015.09.063

[42] Khan SA, Shahid S, Hanif S, Almoallim HS, Alharbi SA, Sellami H. Green synthesis of chromium oxide nanoparticles for antibacterial, antioxidant anticancer, and biocompatibility activities. International Journal of Molecular Sciences. 2021;**22**:502. DOI: 10.3390/ijms22020502

[43] Khani R, Roostaei B, Bagherzade G, Moudi M. Green synthesis of copper nanoparticles by fruit extract of Ziziphus spina-christi (L.) Willd.: Application for adsorption of triphenylmethane dye and antibacterial assay. Journal of Molecular Liquids. 2018;**255**:541-549. DOI: 10.1016/j.molliq.2018.02.010

[44] Qamar H, Rehman S, Chauhan DK, Tiwari AK, Upmanyu V. Green synthesis, characterization and antimicrobial activity of copper oxide nanomaterial derived from Momordica charantia. International Journal of Nanomedicine. 2020;**15**:2541-2553. DOI: 10.2147/IJN.S240232

[45] Arunachalam K, Annamalai S, Hari S. One-step green synthesis and characterization of leaf extract-mediated biocompatible silver and gold nanoparticles from Memecylon umbellatum. International Journal of Nanomedicine. 2013;**8**:1307. DOI: 10.2147/IJN.S36670

[46] Khan SSSA, Lee C-S, Khan SA, Shahid S, Lee C-S, Khan SSSA, et al. Green synthesis of gold and silver nanoparticles using leaf extract of Clerodendrum inerme; characterization, antimicrobial, and antioxidant activities. Biomolecules. 2020;**10**:835. DOI: 10.3390/biom10060835

[47] Barai AC, Paul K, Dey A, Manna S, Roy S, Bag BG. Green synthesis of

Nerium oleander-conjugated gold nanoparticles and study of its in vitro anticancer activity on MCF-7 cell lines and catalytic activity. Nano Convergence. 2018;**5**:10. DOI: 10.1186/s40580-018- 0142-5

[48] Ghosh S, Patil S, Ahire M, Kitture R, Gurav DD, Jabgunde AM, et al. Gnidia glauca flower extract mediated synthesis of gold nanoparticles and evaluation of its chemocatalytic potential. Journal of Nanobiotechnology. 2012;**10**:17. DOI: 10.1186/1477-3155-10-17

[49] Baek K, Patra JK, Baek K-H. Novel green synthesis of gold nanoparticles using Citrullus lanatus rind and investigation of proteasome inhibitory activity, antibacterial, and antioxidant potential. International Journal of Nanomedicine. 2015;**10**:7253-7264. DOI: 10.2147/IJN.S95483

[50] Yusefi M, Shameli K, Yee OS, Teow S-Y, Hedayatnasab Z, Jahangirian H. Green synthesis of Fe_3O_4 nanoparticles stabilized by a *Garcinia mangostana* fruit peel extract for hyperthermia and anticancer activities. International Journal of Nanomedicine. 2021;**16**:2515-2532. DOI: 10.2147/IJN.S284134

[51] Cai W, Weng X, Chen Z. Highly efficient removal of antibiotic rifampicin from aqueous solution using green synthesis of recyclable nano-Fe3O4. Environmental Pollution. 2019;**247**: 839-846. DOI: 10.1016/j.envpol.2019.01.108

[52] Zargar M, Hamid AA, Bakar FA, Shamsudin MN, Shameli K, Jahanshiri F. Green synthesis and antibacterial effect of silver nanoparticles using *Vitex negundo* L. Molecules. 2011;**16**:6667-6676. DOI: 10.3390/molecules16086667

[53] Arokiyaraj S, Valan Arasu M, Vincent S, Oh Y-K, Kim KH, Choi K-C, et al. Rapid green synthesis of silver nanoparticles from Chrysanthemum indicum L and its antibacterial and cytotoxic effects: An in vitro study. International Journal of Nanomedicine. 2014;**9**:379. DOI: 10.2147/IJN.S53546

[54] Oves M, Rauf MA, Aslam M, Qari HA, Sonbol H, Ahmad I, et al. Green synthesis of silver nanoparticles by Conocarpus Lancifolius plant extract and their antimicrobial and anticancer activities. Saudi Journal of Biological Sciences. 2021;**29**:460-471. DOI: 10.1016/j.sjbs.2021.09.007

[55] Alkhalaf MI, Hussein RH, Hamza A, Maha RHH, Alkhalaf I, Hamza A. Green synthesis of silver nanoparticles by Nigella sativa extract alleviates diabetic neuropathy through anti-inflammatory and antioxidant effects, Saudi. Journal of Biological Sciences. 2020;**27**: 2410-2419. DOI: 10.1016/j.sjbs.2020.05.005

[56] Kambale EK, Nkanga CI, Mutonkole B-PI, Bapolisi AM, Tassa DO, Lies-se J-MI. Green synthesis of antimicrobial silver nanoparticles using aqueous leaf extracts from three Congolese plant species (*Brillantaisia patula, Crossopteryx febrifuga and Senna siamea*). Heliyon. 2020;**6**:e04493. DOI: 10.1016/j.heliyon.2020.e04493

[57] Muhammad Irfan HM, Ismail H, Irfan M, Munir H, Ismail H, Muhammad Irfan HM, et al. Moringa oleifera gum based silver and zinc oxide nanoparticles: Green synthesis, characterization and their antibacterial potential against MRSA. Biomaterials Research. 2021;**25**:17. DOI: 10.1186/s40824-021-00219-5

[58] Jayachandran A, TR A, Nair AS. Green synthesis and characterization of

zinc oxide nanoparticles using *Cayratia pedata* leaf extract. Biochemistry and Biophysics Reports. 2021;**26**:100995. DOI: 10.1016/j.bbrep.2021.100995

[59] Kaliraj L, Ahn JC, Rupa EJ, Abid S, Lu J, Yang DC. Synthesis of panos extract mediated ZnO nano-flowers as photocatalyst for industrial dye degradation by UV illumination. Journal of Photochemistry and Photobiology B: Biology. 2019;**199**:111588. DOI: 10.1016/j.jphotobiol.2019.111588

[60] Madan HRR, Sharma SCC, Udayabhanu D, Suresh YSS, Vidya H, Nagabhushana H, et al. Facile green fabrication of nanostructure ZnO plates, bullets, flower, prismatic tip, closed pine cone: Their antibacterial, antioxidant, photoluminescent and photocatalytic properties. Spectrochimica Acta Part A: Molecular and Biomolecular Spectroscopy. 2016;**152**:404-416. DOI: 10.1016/j.saa.2015.07.067

[61] Huq MA. Green synthesis of silver nanoparticles using Pseudoduganella eburnea MAHUQ-39 and their antimicrobial mechanisms investigation against drug resistant human pathogens. International Journal of Molecular Sciences. 2020;**21**:1510. DOI: 10.3390/ijms21041510

[62] Van Tran T, Nguyen DTC, Kumar PS, Din ATM, Jalil AA, Vo D-VN, et al. Green synthesis of ZrO_2 nanoparticles and nanocomposites for biomedical and environmental applications: A review. Environmental Chemistry Letters. 2022;**20**:1-23. DOI: 10.1007/s10311-021-01367-9

[63] Feng Z, Adolfsson KH, Xu Y, Fang H, Hakkarainen M, Wu M. Carbon dot/polymer nanocomposites: From green synthesis to energy, environmental and biomedical applications. Sustainable Materials and Technologies. 2021;**29**:e00304. DOI: 10.1016/j.susmat.2021.e00304

[64] Ganguly S, Das P, Bose M, Das TK, Mondal S, Das AK, et al. Sonochemical green reduction to prepare Ag nanoparticles decorated graphene sheets for catalytic performance and antibacterial application. Ultrasonics Sonochemistry. 2017;**39**:577-588. DOI: 10.1016/j.ultsonch.2017.05.005

[65] Kaushik M, Fraschini C, Chauve G, Putaux JL, Moores A. Transmission electron microscopy for the characterization of cellulose nanocrystals. In: The Transmission Electron Microscope - Theory and Applications [Internet]. London, UK: InTech; 2015. DOI: 10.5772/60985

[66] Anjum DH. Characterization of nanomaterials with transmission electron microscopy. IOP Conference Series: Materials Science and Engineering. 2016;**146**:012001. DOI: 10.1088/1757-899X/146/1/012001

[67] Franken LE, Boekema EJ, Stuart MCA. Transmission electron microscopy as a tool for the characterization of soft materials: Application and interpretation. Advancement of Science. 2017;**4**: 1600476. DOI: 10.1002/advs.201600476

[68] Mokobi F. Transmission Electron Microscope (TEM)- Definition, Principle, Images [Internet]. London, UK: Available from: https://microbenotes.com/transmission-electron-microscope-tem/ [Accessed: August 3, 2023]

[69] Dutrow BL, Clark CM. X-ray powder diffraction (XRD). In: Integrating Research and Education [Internet]. Available from: https://serc.carleton.edu/research_education/geochemsheets/techniques/XRD.html. [Accessed: August 7, 2023]

[70] Malvern Panalytical. X-ray Diffraction (XRD): A non-destructive technique to study all types of material [Internet]. Available from: https://www.malvernpanalytical.com/en/products/technology/xray-analysis/x-ray-diffraction

[71] Nanakoudis A. What is SEM? Scanning electron microscopy explained. In: Thermo Fisher Scientific [Internet]. 2019. Available from: https://www.thermofisher.com/blog/author/antonisnanakoudis/ [Accessed August 7, 2023]

[72] Joy DC, Ford BJ, Bradbury S. Scanning Electron Microscope, Encycl. Br. 2023 [Internet]. Available from: https://www.britannica.com/technology/scanning-electron-microscope [Accessed: August 7, 2023]

[73] S. Swapp, Scanning Electron Microscopy (SEM), Integr. Res. Educ. [Internet]. Available from: https://serc.carleton.edu/research_education/geochemsheets/techniques/SEM.html [Accessed: August 7, 2023]

[74] Pan Y, Blum AS, Simine L, Mauzeroll J. Nanometals templated by tobacco mosaic virus coat protein with enhanced catalytic activity. Applied Catalysis B: Environmental. 2021;**298**: 120540. DOI: 10.1016/j.apcatb.2021.120540

[75] Zhang Y, Tian J, Li H, Wang L, Qin X, Asiri AM, et al. Biomolecule-assisted, environmentally friendly, one-pot synthesis of CuS/reduced graphene oxide nanocomposites with enhanced photocatalytic performance. Langmuir. 2012;**28**:12893-12900. DOI: 10.1021/la303049w

[76] Maleki A, Rahimi J, Demchuk OM, Wilczewska AZ, Jasiński R. Green in water sonochemical synthesis of tetrazolopyrimidine derivatives by a novel core-shell magnetic nanostructure catalyst. Ultrasonics Sonochemistry. 2018;**43**:262-271. DOI: 10.1016/j.ultsonch.2017.12.047

[77] Nasrollahzadeh M, Bagherzadeh M, Karimi H. Preparation, characterization and catalytic activity of CoFe2O4 nanoparticles as a magnetically recoverable catalyst for selective oxidation of benzyl alcohol to benzaldehyde and reduction of organic dyes. Journal of Colloid and Interface Science. 2016;**465**:271-278. DOI: 10.1016/j.jcis.2015.11.074

[78] Chng LL, Erathodiyil N, Ying JY. Nanostructured catalysts for organic transformations. Accounts of Chemical Research. 2013;**46**:1825-1837. DOI: 10.1021/ar300197s

[79] Liu M, Peng T, Li H, Zhao L, Sang Y, Feng Q, et al. Photoresponsive nanostructure assisted green synthesis of organics and polymers. Applied Catalysis B: Environmental. 2019;**249**:172-210. DOI: 10.1016/j.apcatb.2019.02.071

[80] Fattahimoghaddam H, Mahvelati-Shamsabadi T, Lee B-K. Efficient Photodegradation of Rhodamine B and tetracycline over robust and green g-C3N4 nanostructures: Supramolecular design. Journal of Hazardous Materials. 2021;**403**:123703. DOI: 10.1016/j.jhazmat.2020.123703

[81] Muthoosamy K, Geetha Bai R, Abubakar IB, Sudheer SM, Lim HN, Loh H-S, et al. Exceedingly biocompatible and thin-layered reduced graphene oxide nanosheets using an eco-friendly mushroom extract strategy. International Journal of Nanomedicine. 2015;**10**: 1505-1519. DOI: 10.2147/IJN.S75213

[82] Pooresmaeil M, Behzadi Nia S, Namazi H. Green encapsulation of LDH

(Zn/Al)-5-Fu with carboxymethyl cellulose biopolymer; new nanovehicle for oral colorectal cancer treatment. International Journal of Biological Macromolecules. 2019;**139**:994-1001. DOI: 10.1016/j.ijbiomac.2019.08.060

[83] Rabiee N, Fatahi Y, Asadnia M, Daneshgar H, Kiani M, Ghadiri AM, et al. Green porous benzamide-like nanomembranes for hazardous cations detection, separation, and concentration adjustment. Journal of Hazardous Materials. 2022;**423**:127130. DOI: 10.1016/j.jhazmat.2021.127130

[84] Iravani S, Varma RS, Varma RS, Varma RS. Green synthesis, biomedical and biotechnological applications of carbon and graphene quantum dots. A review. Environmental Chemistry Letters. 2020;**18**:1-25. DOI: 10.1007/s10311-020-00984-0

[85] Akintelu SA, Folorunso AS, Folorunso FA, Oyebamiji AK. Green synthesis of copper oxide nanoparticles for biomedical application and environmental remediation. Heliyon. 2020;**6**:e04508. DOI: 10.1016/j.heliyon.2020.e04508

[86] Gurunathan S, Raman J, Abd Malek SN, John PA, Vikineswary S, Sabaratnam V, et al. Green synthesis of silver nanoparticles using Ganoderma neo-japonicum Imazeki: A potential cytotoxic agent against breast cancer cells. International Journal of Nanomedicine. 2013;**8**:4399-4413. DOI: 10.2147/IJN.S51881

[87] Tarannum N, Divya D, Gautam YK. Facile green synthesis and applications of silver nanoparticles: A state-of-the-art review. RSC Advances. 2019;**9**: 34926-34948. DOI: 10.1039/C9RA04164H

[88] Bolade OP, Williams AB, Benson NU. Green synthesis of iron-based nanomaterials for environmental remediation: A review. Environmental Nanotechnology, Monitoring & Management. 2020;**13**:100279. DOI: 10.1016/j.enmm.2019.100279

[89] Shah IH, Manzoor MA, Sabir IA, Ashraf M, Gulzar S, Chang L, et al. A green and environmental sustainable approach to synthesis the Mn oxide nanomaterial from Punica granatum leaf extracts and its in vitro biological applications. Environmental Monitoring and Assessment. 2022;**194**:921. DOI: 10.1007/s10661-022-10606-7

[90] Rónavári A, Igaz N, Adamecz DI, Szerencsés B, Molnar C, Kónya Z, et al. Green silver and gold nanoparticles: Biological synthesis approaches and potentials for biomedical applications. Molecules. 2021;**26**:844. DOI: 10.3390/molecules26040844

[91] Hatamie A, Marahel F, Sharifat A. Green synthesis of graphitic carbon nitride nanosheet (g-C3N4) and using it as a label-free fluorosensor for detection of metronidazole via quenching of the fluorescence. Talanta. 2018;**176**:518-525. DOI: 10.1016/j.talanta.2017.08.059

[92] Nasrollahzadeh M, Sajjadi M, Iravani S, Varma RS. Green-synthesized nanocatalysts and nanomaterials for water treatment: Current challenges and future perspectives. Journal of Hazardous Materials. 2021;**401**:123401. DOI: 10.1016/j.jhazmat.2020.123401

[93] Sadia Saif AT, Chen Y, Saif S, Tahir A, Chen Y, Sadia Saif AT, et al. Green Synthesis of Iron Nanoparticles and Their Environmental Applications and Implications, Nanomaterials. 2016;**6**: 209. DOI: 10.3390/nano6110209

[94] Khan F, Shahid A, Zhu H, Wang N, Javed MR, Ahmad N, et al. Prospects of algae-based green synthesis of

nanoparticles for environmental applications. Chemosphere. 2022;**293**: 133571. DOI: 10.1016/j. chemosphere.2022.133571

[95] Bahrulolum H, Nooraei S, Javanshir N, Tarrahimofrad H, Mirbagheri VS, Easton AJ, et al. Green synthesis of metal nanoparticles using microorganisms and their application in the agrifood sector. Journal of Nanobiotechnology. 2021;**19**:86. DOI: 10.1186/s12951-021-00834-3

[96] Rohimi NF, Yaakob MNA, Roslan R, Salim N, Mustapha SNH, Hasbi Ab. Rahim M, et al. Structural and thermal analysis of bio-based polybenzoxazine derived from liquefied empty fruit bunch (EFB) via solventless method. Materials Today Proceedings. 2022;**51**: 1367-1371. DOI: 10.1016/j. matpr.2021.11.410

[97] Bartolucci C, Antonacci A, Arduini F, Moscone D, Fraceto L, Campos E, et al. Green nanomaterials fostering agrifood sustainability. TrAC Trends in Analytical Chemistry. 2020; **125**:115840. DOI: 10.1016/j. trac.2020.115840

[98] Zhu C, Guo S, Fang Y, Dong S. Reducing sugar: New functional molecules for the green synthesis of graphene Nanosheets. ACS Nano. 2010; **4**:2429-2437. DOI: 10.1021/nn1002387

[99] Xia Y, Xiao Z, Dou X, Huang H, Lu X, Yan R, et al. Green and facile fabrication of hollow porous MnO/C microspheres from microalgaes for lithium-ion batteries. ACS Nano. 2013;**7**:7083-7092. DOI: 10.1021/ nn4023894

[100] Zinatloo-Ajabshir S, Ghasemian N, Mousavi-Kamazani M, Salavati-Niasari M. Effect of zirconia on improving NOx reduction efficiency of Nd2Zr2O7 nanostructure fabricated by a new, facile and green sonochemical approach. Ultrasonics Sonochemistry. 2021;**71**: 105376. DOI: 10.1016/j.ultsonch. 2020.105376

[101] Kalambate PK, Rao Z, Dhanjai J, Wu Y, Shen RB, Huang Y. Electrochemical (bio) sensors go green. Biosensors & Bioelectronics. 2020;**163**: 112270. DOI: 10.1016/j.bios.2020.112270

[102] Deng H-H, Wang F-F, Shi X-Q, Peng H-P, Liu A-L, Xia X-H, et al. Water-soluble gold nanoclusters prepared by protein-ligand interaction as fluorescent probe for real-time assay of pyrophosphatase activity. Biosensors & Bioelectronics. 2016;**83**:1-8. DOI: 10.1016/j.bios.2016.04.031

[103] Rawtani D, Rao PK, Hussain CM. Recent advances in analytical, bioanalytical and miscellaneous applications of green nanomaterial. TrAC Trends in Analytical Chemistry. 2020;**133**:116109. DOI: 10.1016/ j.trac.2020.116109

[104] Nadeem M, Khan R, Afridi K, Nadhman A, Ullah S, Faisal S. Green synthesis of cerium oxide nanoparticles (CeO_2 NPs) and their antimicrobial applications: A review. International Journal of Nanomedicine. 2020;**15**:5951-5961. DOI: 10.2147/IJN.S255784

[105] Padil VVT, Černík M, Thekkae Padil VV. Green synthesis of copper oxide nanoparticles using gum karaya as a biotemplate and their antibacterial application. International Journal of Nanomedicine. 2013;**8**:889-898. DOI: 10.2147/IJN.S40599

[106] Tippayawat P, Phromviyo N, Boueroy P, Chompoosor A. Green synthesis of silver nanoparticles in aloe vera plant extract prepared by a

hydrothermal method and their synergistic antibacterial activity. PeerJ. 2016;**4**:e2589. DOI: 10.7717/peerj.2589

[107] Dikshit P, Kumar J, Das A, Sadhu S, Sharma S, Singh S, et al. Green synthesis of metallic nanoparticles: Applications and limitations. Catalysts. 2021;**11**:902. DOI: 10.3390/catal11080902

Chapter 3

Recent Advances in Bio-Derived Nanomaterials: Green Synthesis of Silica

Ntalane Sello Seroka and Lindiwe Khotseng

Abstract

Silica molecules present in commercial objects can pose a hazard to human health, which is why the environmentally friendly synthesis of silica has been intensively researched in the recent decades. This chapter describes the synthesis of silica from sugarcane bagasse waste and its physical and chemical properties for potential use in eco-friendly applications. Sugarcane bagasse was burned to produce ash, which was then calcined in a 700°C kiln before being treated with citric acid to remove silica from the ash. X-ray fluorescence (XRF) analysis showed that after the acid treatment, 78–79% of the silica was produced and strong peaks were observed in the X-ray diffraction spectra (XRD) at 2Θ = 28 (degree) and an average diameter of 28 nm for 1-HDTA and 30 nm for TPAH, determined by the Scherrer equation. Fourier transform infrared spectroscopy (FTIR) spectra also confirms the presence of synthesized silica. In addition, the shape of the particles was analyzed by TEM and SEM images and it is found that synthesized silica had a spongy shape with irregular sizes ranging from 25 to 50 nm. Overall, the studies show that organic bases are capable of synthesizing silica with application-specific properties from agricultural waste using green chemistry.

Keywords: green synthesis, biomass, nanomaterials, agricultural waste, environmentally friendly

1. Introduction

The synthesis of various nanomaterials for the development of nanoelectronics has sparked a great deal of interest in nanotechnology as a dependable, sustainable, and green process known as "green synthesis." Green synthesis is essentially seen as a significant way to mitigate the issues with traditional methods for creating nanoparticles for use in functional nanomaterials and nanoelectrodes. The four most important variables in the synthesis of nanoparticles according to the green protocol are known to be the selection of an environmentally friendly solvent, a source of nanomaterials, a reducing agent, and a safe material for stabilization [1, 2].

Green chemistry strives to reduce the negative impacts of nanomaterial manufacturing and application by delivering risk-free nanotechnology. Green synthesis is a breakthrough in a variety of new technologies, including energy-saving procedures, large-scale manufacturing of nanoparticles, biological systems, and, finally, an eco-friendly method that does not use poisonous chemicals. A plethora of items have hit the market in recent years, with a select few being integrated into cosmetics, care products, and clothes merely because they are safe to use [2].

The nanomaterials synthesis from agricultural waste is thought to be environmentally friendly. The disposal of sugarcane bagasse attracts a large number of insects that are dangerous to human health and spread a wide range of diseases. This feedstock is used for the production of sugar and ethanol and the residue from the process is disposed as bagasse. The agricultural waste is disposed of in landfills, the majority of agricultural operations produce waste that is produced in large quantities in many countries and poses serious environmental issues [3].

Due to anaerobic decay, organic wastes produce carbon dioxide (CO_2), nitrous oxide (N_2O), and primarily methane (CH_4) in landfills, all of which actively contribute to greenhouse gas emissions. Furthermore, during agricultural production, greenhouse gases primarily CH_4 and N_2O are emitted, with only a trace amount of CO_2 emitted. Reusing agricultural solid waste is regarded as an important waste reduction strategy that meets the requirements of a comprehensive sustainable waste management system; reuse of these wastes can actively contribute to the development of new green technologies, bio-energy generation, and bio-conversion to nanomaterials [3, 4].

Numerous types of nanomaterials have been extracted from agricultural wastes using biogenic synthesis; these synthesis methods are an appealing alternative to conventional synthesis methods because they are green and environmentally friendly. Nanotechnology has emerged as one of the most important branches of science, with numerous industrial applications ranging from the conversion of biomass into valuable materials in niche applications to the production of new materials. The green synthesis of silver nanoparticles (AgNPs) is an excellent example of this; this method employs biomolecules found in plants and agricultural wastes that act as reducing and stabilizing agents. Other noteworthy biogenic processes include the environmentally friendly production of biodegradable polyurethane using castor oil and the production of copper-doped zinc oxide (ZnO) nanoparticles from *Synadenium grantii* leaf extract. Additionally, silicon dioxide (SiO_2), also known as silica, can be extracted from agricultural waste like sugarcane bagasse ash (SCBA) and rice husk (RH) using a sustainable-green synthesis method [5–7].

Plants are viewed as low-cost, non-conservative natural chemical factories. Silver (Ag) can be biosynthesized using waste materials such as safflower (*Carthmus tinctorius* L.), a well-known crop from the northwest of Mexico, according to a study by Rodrguez-Félix, F. et al. [5]. Concurrent with efforts to reduce environmental pollution, industrial agro-waste and food research on the effectiveness and environmentally friendly qualities of phytochemicals from safflower by-products using the ultra-high-performance liquid chromatography–diode array detector–tandem mass spectrometry (UPLC-DAD-MS) method was conducted. The researchers also discussed the quality and development of food packaging made possible by zein films composited with no-ultrafiltrated and ultrafiltrated betalains extract(s) from beetroot (Beta vulgrains) bagasse. It is now possible to synthesize highly valuable silver nanoparticles while also promoting sustainability through bio-inspired methods. Ag that is environmentally friendly and has the potential to be used in the health sector due to its antimicrobial and anticancer properties can be made using extracts and byproducts from agri-food waste [5, 7, 8].

Green synthesis also has the advantages of a feasible methodology, non-hazardous and viable procedures, and a wide range of applications in nanotechnology, biomedicine, and nano-optoelectronics. Because it is chemically inert and has a high melting point, silica is widely used in industry. This makes it highly functional with control over specific properties. The green synthesis of silica appears to be an important field of study with a lot of future growth potential.

2. Materials and methods

The chemicals used in the preparation were purchased from Sigma-Aldrich and were not purified further. Sigma-Aldrich (St. Louis, MO, USA) provided (1-Hexadecyl) trimethyl-ammonium bromide (98%), citric acid (99.5%), and tetrapropylammonium hydroxide (1.0 M) in H_2O. The Sugar Illovo South Africa Company (KwaZulu-Natal, South Africa) supplied the sugarcane. Sugarcane bagasse was soaked in double-deionized water for 24 hours in a typical preparation procedure to remove dust and soil particles. The soaked sugarcane was then oven-dried for 6 hours. The sugarcane bagasse was then burned in the open air to produce black sugarcane bagasse ash (SCBA) **Figure 1**. Deionized water from the Milli-Q water purification system (Millipore, Bedford, MA, USA) was used in the synthesis.

The bagasse ash was mixed with citric acid in a 250 ml and refluxed while stirring at 250 revolutions per minute (rpm) at 90°C for 2 hours, as shown in **Figure 1**. After that, the ash was washed twice with deionized water and decanted until the pH of the supernatant reached 6.5. The resulting ash was then dried in a 40°C oven overnight before being ground into a fine powder with a mortar and pestle. Finally, the acid leached bagasse is calcined in a furnace for four hours at 700°C.

$$(SiO_2)_{48} + [N(C_3H_7)_4]OH \rightarrow [N(C_3H_7)_4]OH(SiO_2)_{48} \tag{1}$$

$$[N(C_3H_7)_4]OH(SiO_2)_{48} + \Delta \rightarrow (SiO_2)_x(s)\downarrow + (CO)_x(g)\uparrow + (H)_x(g)\uparrow + (NO)_x(g) \tag{2}$$

In a typical hydrothermal reaction, two mL of tetrapropylammonium hydroxide were added to the orange-brown multicrystalline product shown in Eq. (1). The solution was then hydrothermally heated in a muffle furnace for 2 hours at 200°C, followed by 1 hour at 600°C in air to burn away the excess carbon, nitrogen, and hydrogen, as shown in Eq. (2). The resultant compound oxidizes C, H, and N to produce oxides and decomposes. As a result, 1-HDTA silica and TPAH-silica nanopowders were synthesized.

3. Results and discussion

The chemical composition of the SCBA was determined using X-ray fluorescence spectroscopy (XRF, Malvern Panalytical, Malvern, UK) for the major oxides listed in **Table 1**. According to the XRF analysis, the leaching process was efficient because the amount of silica was significantly reported to be 79.40 wt% and 78.79 wt% for the two organic acids, respectively. The leaching acid used in the pretreatment step was L-cysteine hydrochloride monohydrate. **Table 1** also shows the loss on ignition. The significant improvement in silica morphology in ash powders is due to the selective removal of synthesis residues using L-cysteine hydrochloride monohydrate, essentially reducing metallic impurities in the sample. Following acid treatment, the sample

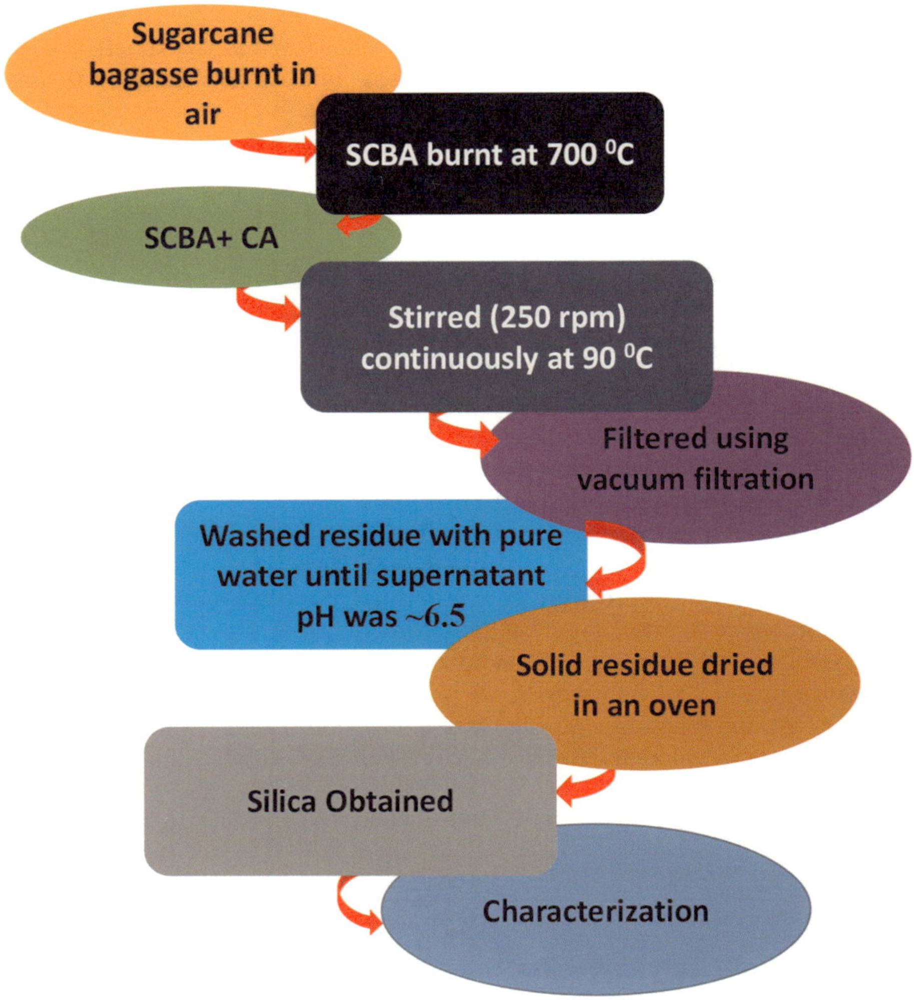

Figure 1.
Schematic illustration for the production of silica from sugarcane bagasse ash.

was calcined in a muffle furnace, and silica was extracted using a hydro-thermal method with tetrapropylammonium hydroxide, minimizing other metal oxide impurities.

3.1 Green silica synthesis

Green and sustainable approaches to obtaining advanced materials for industrial use can help avoid complex procedures while also reducing environmental toxins. Green synthesis methods have been investigated for the extraction of silica from various agricultural biomass sources. Because organic acids, alkalis, and solvents can decompose more complex-structured substrates, silica extraction from biomass is one of the most promising methods for green synthesis. One of the most promising methods for the green synthesis of silica and silicon nanoparticles is the chemical and thermal treatment of biomass by replacing toxic reactants and decreasing temperatures [9, 10].

Component	L-cys (wt%)	CA (wt%)
SiO_2	79.40	78.79
TiO_2	1.03	0.70
Al_2O_3	8.45	8.25
Fe_2O_3	2.39	3.77
MnO	0.04	0.04
MgO	0.61	0.72
Cao	1.06	1.26
Na_2O	1.06	1.01
K_2O	3.86	3.66
P_2O_5	0.22	0.34
SO_3	0.03	0.01
Cr_2O_3	0.04	0.02
NiO	0.02	0/01
H_2O	0.04	−0.13
LOI	1.18	1.50

Table 1.
Chemical composition of acid-leached SCBA from XRF [9].

3.2 Characterization

X-ray diffraction (XRD) on a Bruker AXSD8 Advancement equipment (Ithemba Labs, South Africa) with Cu-K1, λ = 154,050 A was used to cross-check the crystallinity identification. The Bragg angle array has a scanning step of 0.035°C and a temperature range of 2Θ = 10–90°C. IR was used to investigate the new surface functions by detecting the functional groups and bonding of elements present in the samples. The examination was performed at room temperature using a PerkinElmer FTIR spectrometer (spectrum 2) with a wavelength range of 400–4000 cm1. The surface morphology of SCBA was examined using scanning electron microscopy (TESCAN, VEGA). Before analysis, the samples were prepared on an aluminum stub and carbon sputtered on a carbon coater. Transmission Electron Microscopy (TEM) was used to study the structural morphology of the as-synthesized nanoparticles in order to assess particle size distribution and form.

Using Scherrer's formula, the average diameter sizes of silica extracted with (1-Hexadecyl) trimethyl-ammonium bromide, (1-HDTA), and Tetrapropylammonium hydroxide (TPAH) were determined to be around 28 nm and 30 nm, respectively. Scherrer's formula was used to estimate crystallite size, as stated in Eq. (3) below:

$$D_p = \frac{K\lambda}{\beta\cos\theta} \tag{3}$$

Dp is the average crystallite size (nm), and K is the Scherrer constant. K values range from 0.68 to 2.08, with K = 0.94 being preferred for spherical crystallites with cubic symmetry and -X-ray wavelength. CuKα the λ = 1.54178 is commonly used for micro XRD. Furthermore, the peak position is identified by the FWHM (full Width at Half Maximum) of the XRD peak calculated from one half of 2.

3.2.1 Powder X-ray diffraction analysis

Powder XRD was used to determine the structural characteristics of the powder. The SCBA was reacted with two distinct organic bases, 1-HDTA and TPAH, resulting in crystalline silica with average diameters of 28 nm for 1-HDTA and 30 nm for TPAH. This study effectively demonstrated the utilization of organic compounds for silica extraction from SCBA via a green synthesis technique [9]. The nature of crystallized phases is determined by measuring the angles of diffraction of X-rays by the solid's crystalline plane. These diffraction angles are connected to the amorphous and crystal lattice properties. **Figure 2a** demonstrates that SCBA is primarily amorphous, with weak intensities of crystalline peaks present in the phase composition.

Diffraction peaks at 2 theta angles of 22, 28, 39, 48, 56, and 64 were observed in the XRD spectra of (b) 1-HDTA silica and (c) TPAH silica, corresponding to the crystalline phases of silica. The diffraction peak at 2 theta angle 28 reveals that the generated silica nanoparticles have a 101 cubic phase crystalline structure. Similarly, Katare, V. D., and Madurwar, M.V. (2017) [11] observed crystalline silica peaks at 2Θ = 26 [12]. This finding confirms that silica nanoparticles were extracted from SCBA in our samples [9, 11].

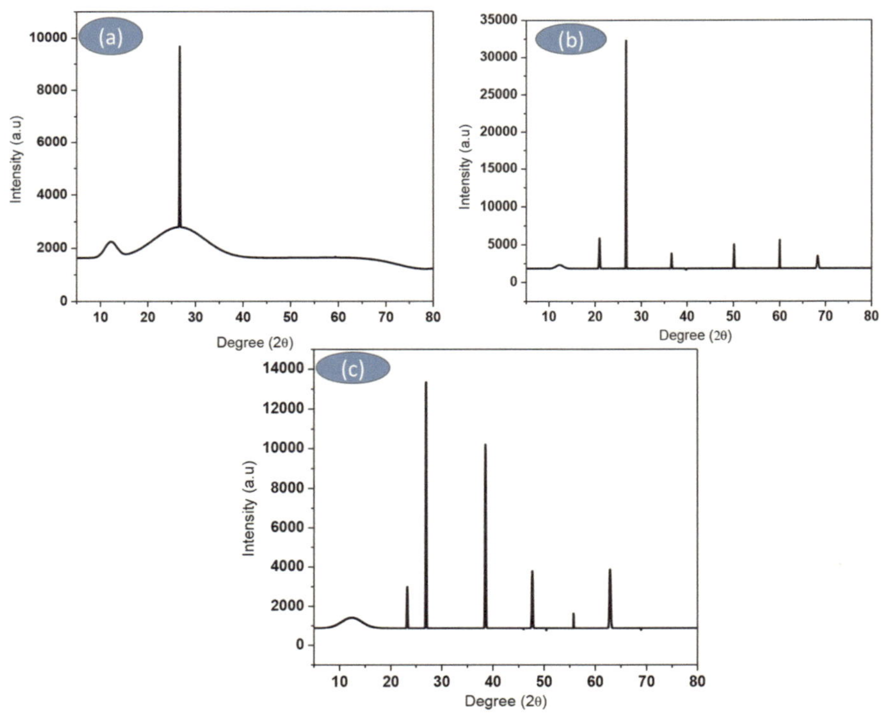

Figure 2.
The XRD patterns for (a) raw bagasse, (b) 1-HDTA silica, and (c) TPAH silica.

3.2.2 FTIR analysis

Figure 3b and **c** mention the characteristic peak number in the IR of the produced nano-silicas, display bands at 461.231 cm^{-1}, 787.381 cm^{-1}, and 1045.99 cm^{-1}, which correspond to the Si~O~Si bending vibration, Si~O~Si stretch vibration, and Si~O~Si stretching vibration, respectively.

The evolution and disappearance of functional groups, as well as the appearance of more prominent groups, demonstrate the incorporation of new functionalities in our samples. The carbonyl group band is visible in **Figure 3a**, but not in the silica spectra of either sample in (b) or (c), confirming the effective conversion of SCBA into silica groups, which are prevalent in the material. The band at 3014.44 cm^{-1} also corresponds to the C-H stretch in (a). SCBA has a weak absorption band for the hydroxyl group (OH) at about 3445.23 cm^{-1} [11].

The nanosilicas produced in (b) and (c) exhibit identical behavior for novel functionality. Despite the impressive results, the silica in **Figure 3c** displayed a generally intense and prominent band at roughly 1060.76 cm^{-1}, which is typical of crystalline silica and confirms the results obtained from XRD diffraction patterns [11].

3.2.3 TEM and SEM analysis

TEM determines the particle size and shape of the materials. **Figure 4a** depicts the intensive particle aggregation in SCBA, as well as the produced silica in (b) and (c). Because the particle size is not consistent, it does not represent single particles but

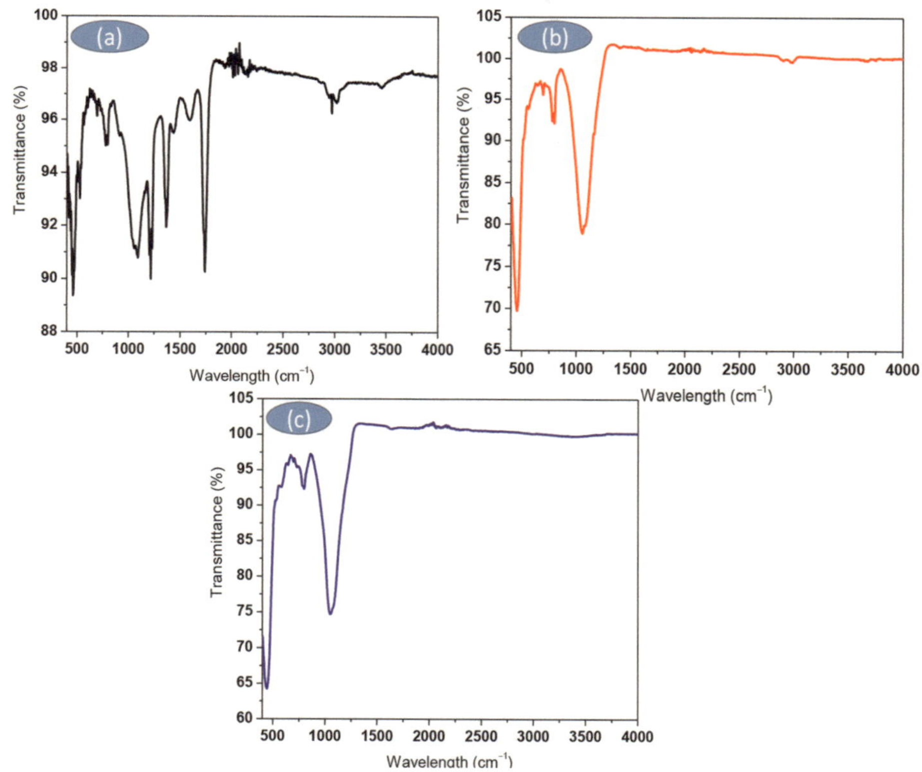

Figure 3.
FTIR spectra of (a) raw SCBA (b) 1-HDTA silica, and (c) TPAH silica.

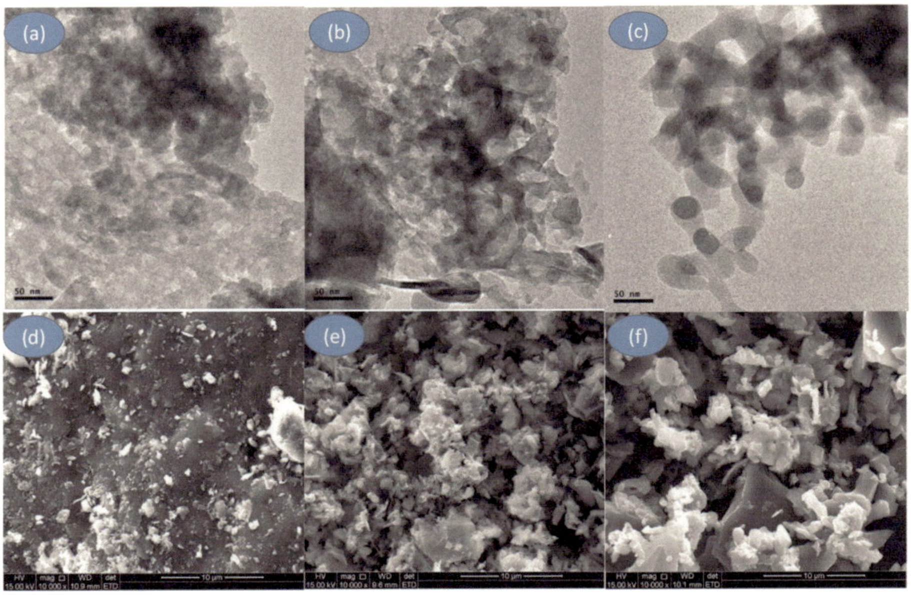

Figure 4.
TEM images of (a) raw SCBA, (b) 1-HDTA-silica, (c) TPAH silica and SEM images of (d) raw SCBA, (e) 1-HDTA and (f) TPAH silica.

rather nanoscale agglomerates (50–100 nm). The silica particles are uniformly dispersed after the TPAH reaction. The particle sizes of the obtained silica and the acid leached using 1-HDTA are distributed randomly. The morphology of the nanosilica derived from the SCBA is identical to that obtained and reported elsewhere [13]. In contrast, agglomerated particles are usually in the nanorange. In terms of size control, this method of green synthesis of nano-silica particles is promising; nonetheless, non-uniform silica nanoparticles were also created.

In all samples linked with the release of organic matter during bagasse burning, the SEM images in **Figure 4d–f** indicate agglomerates of heterogeneous material with irregular forms and high roughness. Although the organic acid leaching process was effective on the surface morphology of the prepared silicas, the morphological differences (f) with fewer aggregates confirm the composition analysis (XRF) discussed earlier, in which pre-treatment with L-cysteine significantly improved silica formation for the acid-leached sample. The acid treatment during the synthesis stage causes particle agglomeration, which is more obvious in the SEM image (c). After TPAH treatment, the SEM picture of the silica sample is clearer, showing that activated carbon lowered the degree of alkalinity for silica, resulting in less agglomeration [13].

3.2.4 Textural analysis

SCBA for nano CA-Silica had BET_{SSA}, pore volume, and pore diameter of 21.6511 m^2/g, 0.04312 cm^3, and 8 nm, respectively, while SCBA for nano L-cys Silica had BET_{SSA}, pore volume, and pore diameter of 116.005 m^2/g, 0.1828 cm^3/g, and 6 nm, respectively. The amount of adsorbed gas grows steadily with increasing P/P_0 ratio at the lower P/P_0 areas for CA-Silica, as shown in **Figure 5a**, whereas it increases dramatically for L-Cys Silica, as shown in **Figure 5b**. As mentioned in, this is commonly attributed to monolayer and multilayer adsorption [14].

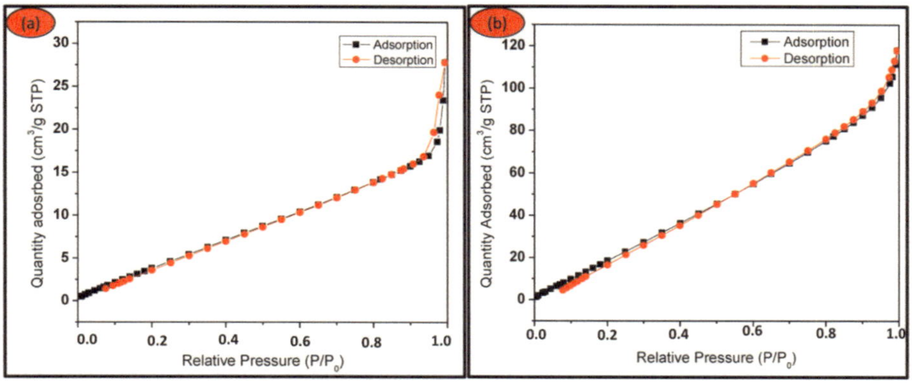

Figure 5.
N_2 adsorption-desorption isotherms for SCBA nano CA-Silica and SCBA nano L-cys silica that is made from SCBA, respectively. Adapted with permission from [9], copyright MDPI (2022).

The hysteresis loop is detected at a range of $0.9 < P/P0 < 1.0$ for the SCBA nanosilica isotherm in **Figure 5a**, which is primarily connected with capillary condensation taking place in the mesopores. Furthermore, the loop on the high P/P_0 side has been linked to big pores. The little loop in **Figure 5b**, detected in a range of $0.2 < P/P_0 < 0.4$, is also caused by capillary condensation within the mesoporous structure; however, no loop is observed at the higher P/P_0 area for this synthesized silica (SCBA nano L-cys Silica) [14, 15].

4. Recommendation

As an alternative to SCBA pretreatment, leaching with an organic acid, specifically citric acid (CA, C6H8O7). This organic acid has the potential to act as both a chelating agent and a mild triprotic acid. CA has the advantage of being an organic and bio-based acid, which has an impact on the materials synthesized and the effluent treatment costs, and it is an eco-friendly leachate when compared to strong mineral acids like HCl and H_2SO_4. Rodrguez-Machn et al. investigated the effect of leaching using CA for the pretreatment of SCBA with varying degrees of thermal degradation and its behavior. As shown in **Table 2**, CA was compared to two well-known conventional leaching agents, HCl and H_2SO_4, in terms of their impact on the chemical, structural, and thermal properties of leached ash. Inorganic element removal ranged from 39.9% to 54.1% depending on leaching duration and temperature, and CA performed similarly to strong mineral acids. When compared to traditional acids for the pretreatment of SCBA, this study shows that CA is a functional leaching agent [16].

Table 3 illustrates the pre-treatment efficiency of HCl and demonstrates significant increase in the silica content and appreciable decrease impurities. To obtain silica with high purity metal oxides which are present as impurities defeats this goal to a certain extent. Thus, purification strategies are required to improve purity, such as thermochemical processes [17, 18].

Organic chemicals and less energy can be used to create green silica nanoparticles, allowing for fine control of particle form, size, and morphology. As a result, new innovative tactics and extra research have been developed to improve

Element	Pristine SCBA	HCl	H_2SO_4	$C_6H_8O_7$
K	1800	17.5	15.7	17
Al	279	110	122	131
Fe	327	110	101	123
Si	8600	3600	4400	4400
Mg	287	25.1	26.4	68.1
Na	32.1	18.6	8.02	<5

Table 2.
Comparison between conventional strong acids (HCl and H_2SO_4) and an organic acid (CA, $C_6H_8O_7$). Inorganic element composition in mg kg^{-1}.

Component	Bagasse ash (Pristine) (wt%)	Bagasse ash (acid-treated) (wt%)
SiO_2	53.10	88.13
MgO	20.72	3.04
P_2O_5	7.36	1.15
SO_3	11.23	4.69
MnO	1.45	0.62
Fe_2O_3	0.78	0.94
K_2O	1.26	0.50
CaO	3.77	0.57
Other	0.36	0.32

Table 3.
SCBA treated with HCl, before and after [17].

existing green technologies and methodology for future silica preparation. Toxic chemicals are utilized in the production of silica from SCBA, which is exceedingly dangerous to both the environment and humans. To begin, 1 M NaOH is used to extract silica in a water bath heated to 90°C, as given in Eq. (4), traditionally known as alkaline treatment. The sample is then filtered and decant supernatant several times until the pH is neutral, and it is dried in an oven. The as-obtained sodium silicate is then immersed in a strong acid 1 M HCl for neutralization reaction in a reflux for an hour at 75°C to make silica gel from sugarcane bagasse, as given in Eq. (5) [12, 19].

$$SiO_2(s) + 2NaOH(aq) \rightarrow Na_2SiO_3(aq) + H_2O(aq) \quad (4)$$

$$Na_2SiO_3(aq) + 2HCl(aq) \rightarrow SiO_2(gel) + 2NaCl(aq) + H_2O \quad (5)$$

Greener and more sustainable synthetic technologies with eco-friendliness, low cost, low energy, low or non-toxicity, and easy procedures are extremely promising and should be prioritized by researchers. The utilization of proteins in the synthesis of multifunctional silica nanoparticles has important economic and environmental implications. To minimize high temperatures, energy, pressure, and the use of toxic and/or hazardous substances and circumstances, sustainable and eco-friendly silicon nanoparticle production technologies with enticing advantages over conventional methods are being developed. Organic acids and bases were used in this study to fulfill the green chemistry goal.

5. Conclusion

The green synthesis of silica nanoparticles from biomass waste (bagasse) is the basis of this research. The use of toxic-free reagents and substances in the green chemistry method for the synthetic route of nanoparticles is gaining traction. When compared to traditional approaches, this approach is seen as more environmentally friendly. It is critical to recognize the non-toxicity of nanoparticles produced using green chemistry. Green chemistry is less expensive, non-toxic, and better for the environment.

The silica nanoparticles were effectively manufactured utilizing the green synthesis approach. Meanwhile, the XRD spectra revealed the primary peaks of silica 2 = 28°, whilst the bands at 1045.99 cm^{-1} on the FTIR spectrum suggested silica production in the samples. Furthermore, approximately 80% of the SiO_2 was removed utilizing the organic acid/base treatment method on the SCBA. The particle size distribution ranged from 25 to 500 nm. The synthesized nanomaterials revealed residue of impurities associated commonly with the green method.

Acknowledgements

This research was funded by National Research Foundation: 138079 and Eskom (South Africa): 2002/015527/0. This research work was supported by the University of the Western Cape and the Sustainable and Renewable Energy Nanomaterials research group.

Author details

Ntalane Sello Seroka* and Lindiwe Khotseng
Faculty of Natural Science, Department of Chemistry, University of the Western Cape, Bellville, South Africa

*Address all correspondence to: 3754640@myuwc.ac.za

References

[1] Duque-Acevedo M, Belmonte-Ureña LJ, Cortés-García FJ, Camacho-Ferre F. Agricultural waste: Review of the evolution, approaches and perspectives on alternative uses. Global Ecology and Conservation. 2020;**22**:00902

[2] Shafey AME. Green synthesis of metal and metal oxide nanoparticles from plant leaf extracts and their applications: A review. Green Processing and Synthesis. 2020;**9**(1):304-339

[3] Danevad D, Carlos-Pinedo S. Exploring interactions between fruit and vegetable production in a greenhouse and an anaerobic digestion plant—Environmental implications. Frontiers in Sustainability. 2021;**2**:111

[4] Mazari SA, Ali E, Abro R, Khan FSA, Ahmed I, Ahmed M, et al. Future challenges-a review. Journal of Environmental Chemical Engineering. 2020;**2020**:105028. DOI: 10.1016/j.jece.2021

[5] Rodríguez-Félix F, Graciano-Verdugo AZ, Moreno-Vásquez MJ, Lagarda-Díaz I, Barreras-Urbina CG, Armenta-Villegas L, et al. Trends in sustainable green synthesis of silver nanoparticles using agri-food waste extracts and their applications in health. Journal of Nanomaterials. 2022;**2022**: 1-37

[6] Zuliani A, Rapisarda M, Chelazzi D, Baglioni P, Rizzarelli P. Synthesis, characterization, and soil burial degradation of biobased polyurethanes. Polymers. 2022;**14**(22):4948

[7] Karthik KV, Raghu AV, Reddy KR, Ravishankar R, Sangeeta M, Shetti NP, et al. Green synthesis of Cu-doped ZnO nanoparticles and its application for the photocatalytic degradation of hazardous organic pollutants. Chemosphere. 2022; **287**:132081

[8] Rodríguez-Félix F, Corte-Tarazón JA, Rochín-Wong S, Fernández-Quiroz JD, Garzón-García AM, Santos-Sauceda I, et al. Physicochemical, structural, mechanical and antioxidant properties of zein films incorporated with no-ultrafiltered and ultrafiltered betalains extract from the beetroot (Beta vulgaris) bagasse with potential application as active food packaging. Journal of Food Engineering. 2022;**334**:111153

[9] Seroka NS, Taziwa R, Khotseng L. Green synthesis of crystalline silica from sugarcane bagasse ash: Physico-chemical properties. Nanomaterials. 2022;**12**(13): 2184

[10] Kalapathy U, Proctor A, Shultz J. Production and properties of flexible sodium silicate films from rice hull ash silica. Bioresource Technology. 2000; **72**(2):99-106

[11] Katare VD, Madurwar MV. Experimental characterization of sugarcane biomass ash–a review. Construction and Building Materials. 2017;**152**:1-15

[12] Seroka NS, Taziwa RT, Khotseng L. Extraction and synthesis of silicon nanoparticles (SiNPs) from sugarcane bagasse ash: A mini-review. Applied Sciences. 2022;**12**(5):2310

[13] Carmona VB, Oliveira RM, Silva WTL, Mattoso LHC, Marconcini JM. Nanosilica from rice husk: Extraction and characterization. Industrial Crops and Products. 2013;**43**:291-296

[14] Shen L, Guo X, Fang X, Wang Z, Chen L. Magnesiothermically reduced diatomaceous earth as a porous silicon

anode material for lithium ion batteries. Journal of Power Sources. 2012;**213**: 229-232

[15] Li M, Dai Y, Ma W, Yang B, Chu Q. Review of new technology for preparing crystalline Silicon solar cell materials by metallurgical method. In: IOP Conference Series: Earth and Environmental Science. Vol. 94, No. 1. Bristol, United Kingdom: IOP Publishing; 2017. p. 012016

[16] Rodríguez-Machín L, Arteaga-Pérez LE, Pérez-Bermúdez RA, Casas-Ledón Y, Prins W, Ronsse F. Effect of citric acid leaching on the demineralization and thermal degradation behavior of sugarcane trash and bagasse. Biomass and Bioenergy. 2018;**108**:371-380

[17] Norsuraya S, Fazlena H, Norhasyimi R. Sugarcane bagasse as a renewable source of silica to synthesize Santa Barbara Amorphous-15 (SBA-15). Procedia Engineering. 2016;**148**:839-846

[18] September LA, Kheswa N, Seroka NS, Khotseng L. Green synthesis of silica and silicon from agricultural residue sugarcane bagasse ash–a mini review. RSC Advances. 2023;**13**(2): 1370-1380

[19] Thuc CNH, Thuc HH. Synthesis of silica nanoparticles from Vietnamese rice husk by sol–gel method. Nanoscale Research Letters. 2013;**8**:1-10

Chapter 4

The Role of Mangroves and Nanomaterials in the Heavy Metals' Decontamination Process

Keyla Soto Hidalgo

Abstract

Green Chemistry aims to make processes more efficient and create products that reduce environmental pollution. This chapter presents the results of the combined use of *Rhizophora mangle* (*R. mangle*) and nano zero-valent iron (nZVI) for cadmium (Cd) remediation in contaminated areas. Here, the phytoremediation process using a 40-ppm cadmium solution in combination with *R. mangle* and nZVI for 3 days was evaluated. We used 20 *R. mangle* samples with and without nZVI to assess the efficiency of the phytoremediation process for the removal of cadmium by inductively coupled plasma (ICP) analytical measurements and confocal imaging. Translocation factors (TFs) were calculated. TF< 1 indicates that *R. mangle* does not act as a hyperaccumulator but as an excluder of metallic Cd. The data obtained in the project suggests that the interaction of nZVI with *R. mangle* is an effective way to enhance the translocation of the metal by its aerial parts without affecting the stomatal opening.

Keywords: heavy metals, nanomaterials, mangroves, green synthesis, phytoremediation

1. Introduction

To maintain the concentration of essential metals within physiological limits and to minimize the harmful effects of nonessential metals, plants have developed complex mechanisms that serve to control the uptake, accumulation, handling, and detoxification of heavy metals. Some of these mechanisms include transport, chelation, compartmentalization, exclusion, and sequestration processes [1]. Of these, chelation constitutes one of the major detoxification mechanisms and involves the participation of high-affinity ligands such as phytochelatins (PCs) and metallothioneins (MTs), two of the best-characterized heavy metal-binding ligands in plant cells [2].

Several studies have reported the benefits of using plants to remove heavy metals from contaminated waters and soils. This process is known as phytoremediation. In the wetlands, plants that naturally thrive in these regions have been found to serve as important tools for heavy metal removal. They are preferred over other bio-agents because they not only uptake nutrients but are also able to adsorb and accumulate heavy metals. Wetland species differ in their ability to uptake and accumulate various metals in their tissues [3]. Some species have been found to have a great ability to accumulate metals in their biomass and are referred to as "hyperaccumulators" [4].

1.1 Mangroves and heavy metals

Mangrove forests are plant associations of great ecological service and economic importance [5, 6]. Mangroves are important wetland plants distributed in the transition zone between land and sea along intertidal coasts in the tropics and subtropics [7]. They form natural barriers with high productivity that protect coastal zones from tropical storms and hurricanes and function as a nursery for numerous species of terrestrial and aquatic fauna [8]. Mangroves have been considered a highly tolerant group of plants because they can survive stressful conditions in tropical and subtropical latitudes [9]. Many mangrove ecosystems located close to urban areas may be impacted by effluents from industrial sources and urban runoff that often contain high concentrations of toxic heavy metals such as cadmium, arsenic, lead, zinc, and nickel, among others [10].

Avicennia spp. and *Rhizophora mangle* are the most tolerant mangrove species. They have been reported to accumulate greater quantities of metals when compared to other mangrove species, even before any visible signs of toxicity appear. Many studies described these plant properties as a complex process of detoxification and tolerance mechanisms to heavy metals that consist of (a) binding to the cell wall and extracellular exudates, (b) reduced uptake or efflux pumping of metals at the plasmatic membrane, (c) detoxification of metals in the apoplasts and chelation of metals in the cytoplasm with various ligands, such as phytochelatins, metallothioneins, metal-binding proteins, and (d) sequestration of metals into the vacuole by tonoplast-located transporters [11, 12]. Nevertheless, these mechanisms in plants are not enough by themselves to substantially remove heavy metals from contaminated sites.

1.2 Nanomaterials for environmental treatment

Nanomaterials iron-based have a great capacity to react, degrade, adsorb, or transform a wide range of contaminants in soils and water bodies. It is used in soil remediation, water treatment or conditioning, permeable reactive barriers, and other applications. Iron ions remain stable in a dry atmosphere and CO_2-free water. Under other conditions, iron oxidizes to Fe^{2+} and Fe^{3+}, forming FeO, Fe_3O_4, and δ-Fe_2O_3, which are irreversible. The ferric form (Fe^{3+}) is very prone to hydrolysis, giving rise to an insoluble iron hydroxide polymer. The predominant species in aqueous solution are Fe^{2+} and Fe^{3+} and the organic ferrous and ferric complexes.

Under aerobic conditions and neutral pH, inorganic iron (III) salts are the most stable species. The synthesis based on the chemical reduction of sodium borohydride for our studies with mangroves has not represented physiological effects on plants. Even so, it is necessary to consider the green synthesis of nanoparticles. Traditional chemical methods used for the synthesis of nano zero-valent iron (nZVI) still require toxicity studies for potentially hazardous by-products. Therefore, alternative methods of low-cost and green synthesis are urgently required. Plant extracts can be used as a highly efficient and environmentally sustainable alternative for the synthesis of nanoparticles.

1.3 Green synthesis of nZVI

Green synthesis for decontamination processes is essential to protect organisms present in ecosystems. On the other hand, the high cost of many techniques and syntheses greatly impedes the process of purification and environmental abatement.

Green synthesis to produce nZVI was first applied by VeruTEK and the US EPA. In this method, plant extracts (coffee, green tea, black tea, lemon, balm, sorghum, bran, grape, etc.) are heated in water to near boiling point to prepare a polyphenol solution [13]. The extract thus produced is separated from the plant debris and mixed with the Fe^{2+} solution. Iron ions in the presence of polyphenols were reduced to nZVI. It must be considered that the main problem when using plants for the synthesis of nanoparticles is the destruction of plants and plant parts. One way around this is to use agricultural residues, for example, Eucalyptus Leaf Fragments or extracts from various residues (peel, albedo, pulp) of lemons, tangerines, etc. Plant extracts reduce metal ions in a shorter time than microorganisms. Depending on the type of plant and the concentration of plant secondary substances, microbial-based nanoparticles are synthesized in minutes to hours [14].

Recently, *Ficus carica* dried fruit, citrus maxima aquatic peel, eucalyptus, and many *Avicennia marina* flower plants and leaves and green tea have been successfully used as green reducing agents for the synthesis of nZVI [15]. The leaf extract of *Cleistocalyx operculatus* has been used widely in medicine and beverages, containing a high proportion of reductants, including polyphenols. This substance can chemically convert Fe ions into iron with zero valency, as described in study [16]. In other studies, the walnut-green shell was used for the synthesis of spherical nZVI particles with a diameter of 18–72 nm [17]. Replacing chemical-reducing agents with plant extracts as reductants reduced operating costs, improved nZVI quality, and improved process performance. However, each plant produces specific extracts with different chemical properties and reducing power [18]. Moreover, the use of this green synthesis for ZVI nanoparticles is not yet widespread worldwide, as the quality of plant extracts depends on local conditions such as climate and soil factors. This chapter aims to show how the use of *R. mangle* and iron nanoparticles could be used as an alternative to the heavy metal decontamination process.

2. Experimental methods results

2.1 Materials

Iron (III) chloride hexahydrate (97%, ACS reagent), sodium borohydride (98.5%, reagent grade), $FeCl_3 \cdot 6H_2O$ (F2877 Sigma Aldrich), $NaBH_4$ (Alfa Aesar), and ethanol 200 proof, 99. All solutions were made using deionized water (18.2 M-cm, Nanopure Diamond, Barnstead).

2.2 Instrumentation and techniques

To analyze the size and morphology of the nanoparticles X-ray diffraction (XRD) and scanning electron microscope (SEM) were employed to determine iron compounds and their respective oxidation states. The phase composition and structures of the nZVI particles were determined by XRD patterns (PANalytical X'Pert Material Research Diffractometer) operating with a Cu Kα radiation (λ = 1.54 Å) source.

To determine the effect of the nZVI nanoparticles in the number of Cd ions in each system and plant after the treatment period time, all the samples were processed and analyzed using ICP-AES Model Perkin Elmer 4300 DV with a detection limit of 0.002 ppm. ICP-AES is one of the most common techniques for elemental analysis. Its high specificity, multi-element capability, and good detection limits result in the use of the technique in a large variety of applications. Three *R. mangle* individuals were selected that were exposed to a system contaminated with 10 ppm cadmium, and another three were exposed to a 1 g/L solution of nZVI only. Translocation Factor (TF) Technique was used to measure the Cd translocation from shoot to root, which is given below:

$$TF = C_{shoot} / C_{root} \quad (1)$$

where C_{shoot} and C_{root} are metals concentration in the shoot ($mg{\bullet}kg^{-1}$) and root of the plant ($mg{\bullet}kg^{-1}$), respectively. A TF value larger than 1 represents the effective translocation of metals to the shoot from the root. To confirm the effect on stomata, a Nikon Eclipse Ti-E inverted fluorescence microscope confocal microscope images were obtained using *R. mangle* leaves from the cadmium solution system and the nZVI and cadmium solution system.

2.3 Sampling and acclimation

The seedlings of *R. mangle* were obtained from Pinones State Forest through direct harvesting during January 2023 (**Figure 1**). This area is the most extensive natural system of mangroves in Puerto Rico. Mangrove plants were first kept in a nursery station under shaded cover using the same substrate of marsh to maintain their natural conditions.

The growth and maintenance of these plants once brought from the wetland was performed in a greenhouse chamber adapted by construed four small chambers that were prepared to carry out different experiments. This chamber from Carolina Biological Company allows to manipulation of environmental conditions to investigate plant growth and development. This strong, long-lasting chamber is made of galvanized, powder-coated steel. Dimensions: 26″ L × 26″ W × 37″ H. Around 4600 lumens of light are provided by four compact fluorescent lamps (CFLs). Greenhouse conditions were 25 degrees Celsius and 60–70% humidity. After an acclimation period of 28 to 30 days, surviving plants are considered fully adapted to their new environmental conditions in the phytonanoremediation system, as seen in **Figure 2**.

The phytonanoremediation system shown in **Figure 2** was adapted using a conceptual design of direct injection in situ. All the systems have water and soil with the highest concentration of Cd to simulate contaminated sites. The purpose of this

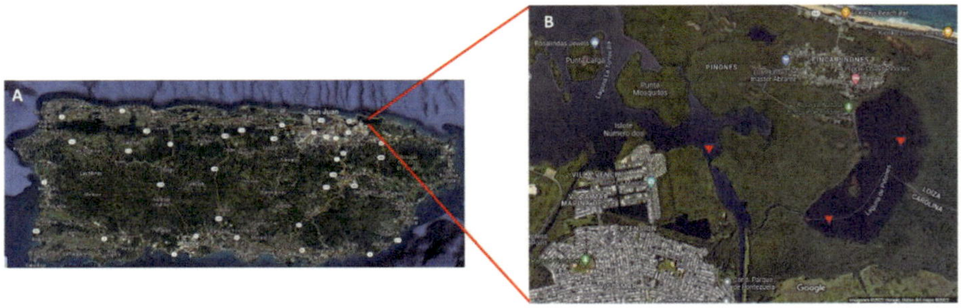

Figure 1.
Puerto Rico map (A), sampling zones and mangrove areas at State Forest Pinones marsh (B).

DOI: http://dx.doi.org/10.5772/intechopen.1002296

Figure 2.
Images of R. mangle obtained from State Forest Pinones and transplanted to a new system for phytonanoremediation system. The experiment employed a total of 20 seedlings. Each one's height and chlorophyll concentration in the leaves were measured.

phytonanoremediation system is to evaluate if the interaction of *R. mangle* with the nZVI in soil and water collectively mitigates and removes heavy metals.

2.4 Synthesis of nZVI

nZVI particles were produced by adding $NaBH_4$ as a reducing agent to a solution of $FeCl_3{\cdot}6H_2O$. The resulting reductive reaction may be given as:

$$Fe^{3+} + 6\,BH^{4-} + 18\,H_2O \rightarrow 2\,FeO(s) + 6\,B(OH)_3 + 21H_2 \qquad (2)$$

To obtain approximately 1.00 g of nZVI, a 0.6 M solution of $FeCl_3$. $6H_2O$ (molar mass 270.30 g/mol) was prepared in 30 mL of ethanol (83% v/v) in nanopure water. The solution was purged with N2 for 30 minutes to remove oxygen before the reaction to avoid the rapid oxidation of iron. Then, the solution was titrated by adding a total of 100 mL of 0.8 M aqueous solution of $NaBH_4$ under a nitrogen atmosphere. After 30 minutes of stirring with a magnetic bar at 600 rpm, the solution was filtered using a 0.22 μm filter paper (Millipore) under a vacuum at 25°C. The sample was rinsed three times with 99.5% absolute ethanol. Various filtered samples, prepared as mentioned, were placed immediately in a vacuum desiccator. Each synthesis of nZVI was prepared to use all the nanoparticles in each treatment.

2.5 Preparation of contaminated solution

Cd solutions were prepared using $Cd(CH_3COO)_2{\cdot}2H_2O$ due to their high solubility in water and ethanol. To represent different samples of wastewater contaminated with Cd solutions of 40 ppm were prepared. The solutions of Cd were treated with 1.0 g/L of nZVI particles for a period of 1 hour. Concentration units were presented as "ppm" keeping with toxicity nomenclature (ppm = mg·L^{-1}). After 1 hour of stirring with a magnetic bar at 600 rpm, the solution was filtered using a 0.22 μm filter paper (Millipore) under vacuum conditions at 25°C. The filtered samples were placed immediately in a desiccator to be analyzed.

In the phytonanoremediation system, two treatments were carried out: the first involved immersing *R. mangle* seedlings in a 40-ppm cadmium solution, and the second involved immersing *R. mangle* seedlings in a 1 g/L nZVI solution near the roots. Chlorophyll content changes if a plant is stressed. A SPAD meter can be used for a rapid determination of the chlorophyll content of individual leaves. An important indicator of plant health is its concentration of chlorophyll, which will be measured using a chlorophyll meter model SPAD (Konica Minolta company) on each plant for a period of 3 days. The chlorophyll meter measures the light transmittance in two wavelength ranges (600 to 700 and 400 to 500 nm) to determine the relative amount of chlorophyll in the leaves. The three treatments included 40- ppm cadmium solution, 1 g/L nZVI solution, and a control group with only water. The chlorophyll levels were measured on three leaves per individual and averaged out in triplicate, resulting in a total of nine leaves analyzed for each treatment.

2.6 Characterization of nZVI

Nanoscale zero-valent iron particles were produced by a chemical reduction technique of $FeCl_3.6H_2O$ using $NaBH_4$ to complete Cd^{2+} fixation from water [19]. In **Figure 3**, X-ray diffraction of nZVI shows a characteristic peak of Fe^0, usually appearing at 2θ = 44.59. The SEM image shows nano spherical iron particles that form chains like aggregates due to their magnetic properties. The diameter of the iron nanoparticles was in the range of 20–100 nm.

The nZVI particles (1 g/L) interact with cadmium ions in the solution for 1 contact hour. Energy dispersive X-ray spectroscopy (EDS) was used to qualitatively determine the morphology, size, and composition of the samples. **Figure 4** shows images of the surface of the ZVI nanoparticles that were taken at different extensions. Simultaneously, the EDS spectrum was obtained in selected areas on the nZVI surface to give information on the chemical composition and surface atomic distributions of

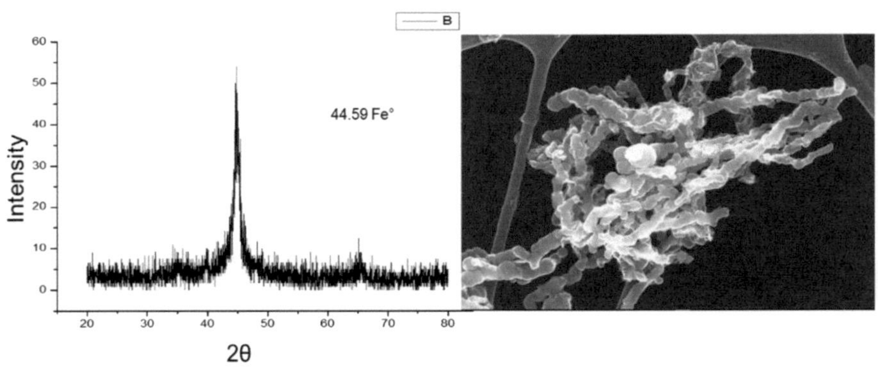

Figure 3.
X-ray diffraction (XRD) pattern and scanning electron microscope (SEM) image of freshly prepared nZVI particles.

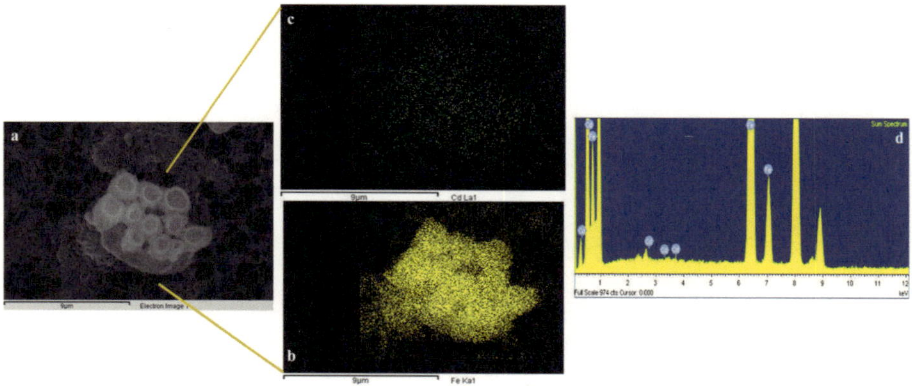

Figure 4.
SEM image of nZVI particles in cadmium solution (a) image of the selected area of nZVI in Cd solution (yellow dots represent Fe and green dots Cd atoms, (b and c respectively)), (d) EDS spectrum of nZVI particles in the Cd solution.

iron and cadmium. The concentration of Cd^{2+} used was 40 ppm, which is within the range of values found in Cd-contaminated areas of Puerto Rico. These results on the distribution of Cd and Fe ions are consistent with our previous studies [20]. When removing cadmium ions, Fe nanoparticles oxidized as FeO and other iron oxides (such as $Fe(OH)_2$) form favorable interactions as cadmium ferrites by the following equation.

$$2Fe^{2+} + Cd^{2+} + 4OH^{-} \rightarrow CdFe_2O_4 + 2H_2 \quad (3)$$

Our studies showed that using *R. mangle* plants with nZVI exposed to cadmium accumulated high concentrations of metals in the roots and aerial organs. The TF was 1.223×10^{-4} for plants exposed for three days to 40 ppm Cd solutions for ICP analysis in the roots, stem, and leaves, and it was 1.420×10^{-2} for plants exposed to nZVI and cadmium solution. Even though these differences are not statistically significant, the TF of *R. mangle* seedlings exposed to nZVI and cadmium exhibits higher translocation mobility. The longer the nanoparticles are exposed, the more efficient the phytonanoremediation process will be.

Results of chlorophyll analyses were measured in *R. mangle* to compare the possible effect of nZVI in plants. Results indicate that the concentration of chlorophyll content of systems does not change (see **Figure 5**). These results confirm that the presence of nZVI can interact with different mangrove species to remove heavy metals without affecting the physiology of plants.

In **Figure 6**, we can show the interaction of nZVI and Cd in the stomata of red mangroves. Images A and B show the stomata open, but image C with the presence of cadmium, directly affects the structure and morphology of the stomata. Confocal microscopy images, taken at low and high magnification, of guard cells after one week of interaction with cadmium and nZVI solutions. The average stomatal openings of control (a), nZVI exposure (b), and Cd solution (c). The stomatal openings in plants exposed to nZVI allow the gas exchange of carbon dioxide and oxygen for

Figure 5.
Variation of chlorophyll content in control, treated with 1 g/L of nZVI and treated with 40 ppm of Cd solution during an exposure period of 3 days. Means ± SD (n = 9 per treatment). When exposed to Cd, the green color of seedling leaves changes due to the decrease in chlorophyll concentration.

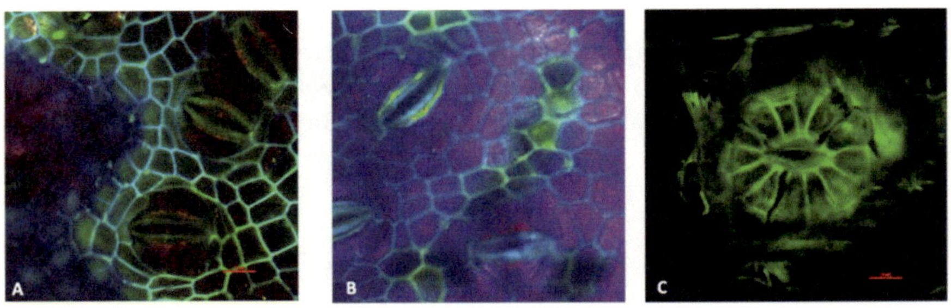

Figure 6.
Confocal microscopy images shows the interaction of nZVI and Cd in the stomata of R. mangle. Images A and B show the stomata open, but image C with the presence of cadmium, directly affects the structure and morphology of the stomata. Red scale bars: 20 μm.

efficient photosynthesis. Our results indicate that nZVI can promote plant growth by increasing their photosynthesis and nutrient accumulation.

3. Role of green chemistry and plants in the removal of heavy metals

One of the main objectives of green chemistry is to develop and produce chemical products and processes that reduce pollution. There are many processes and techniques for the removal of heavy metals, including reverse osmosis, chemical precipitation, ion exchange, adsorption, and solvent extraction [21]. However, these techniques require significant maintenance capacity and cost and are generally not sustainable. For a process to comply with the principles of green chemistry, the degradation of active products must be considered because many chemicals do not fully break down into carbon dioxide, water, and inorganic salts.

Chemical generation may be resistant to environmental biodegradation including microorganisms present. In the environmental area, the use of decontaminants is a critical point of concern since many current chemicals can bioaccumulate and cause toxic effects in organisms. For a compound to be considered biodegradable, its structure, size, and application must be evaluated. Environmental conditions in which they can react, etc. Therefore, the products considered in green chemistry must be biodegradable and nontoxic [22]. The use of plants is an economical and alternative method to synthesize decontamination products for medicines. And it offers one of the greenest ways to overcome costs and toxic effects.

4. Conclusions

The data obtained in these studies suggest that the interaction of nZVI with *R. mangle* is an important way to improve the bioavailability of Cd^{2+} and help mangroves to better translocate them in harvestable portions. The presence of nZVI in *R. mangle* facilitates the entry of ions. The tolerance of *R. mangle* seedlings to high concentrations of Cd may be due to the formation of nontoxic sulfides in the roots or on their surfaces, tissue detoxification, and ion exclusion mechanisms within the roots or inside. These preliminary results, therefore, suggest that Rhizophora may act better as phytoremediators against heavy metals in contaminated areas due to their ability to accumulate at low concentrations in aged leaves. By encapsulating pollutants instead of venting them to the outside, they prevent further pollution of the surrounding ecosystem. The use of red mangroves is important because they are highly resilient ecosystems to the effects of climate change. The approach provides a real alternative to remove human carcinogens such as Cd, As, Pb, and Hg by using affordably synthesized particles at the nanoscale and plant levels.

Acknowledgements

This project was financially supported by NSF-CREST Center for Innovation, Research and Education in Environmental Nanotechnology (CIRE2N) Grant Number NSF-HRD-1736093. Thanks to Sr. Herminio Diaz, officer of the Pinones Forest at Department of Natural and Environmental Resources, for providing the necessary seedlings for the study. Thanks to Bismark Madera for the confocal images at the Neuroimaging and Electrophysiology Facility, University of Puerto Rico.

Conflict of interest

The authors declare no conflict of interest.

Author details

Keyla Soto Hidalgo
University of Puerto Rico, Rio Piedras Campus, United States of America

*Address all correspondence to: keyla.soto@upr.edu

References

[1] Clemens S. Molecular mechanisms of plant metal tolerance and homeostasis. Planta. 2001;**212**:475-486

[2] Hall JL. Cellular mechanisms for heavy metal detoxification and tolerance. Journal of Experimental Botany. 2002;**53**:1-11

[3] Rai UN, Sinha S, Chandra P. Wastewater treatability potential of some aquatic macrophytes: Removal of heavy metals. Ecological Engineering. 1995;**5**(1):5-12

[4] Rascio N, Navari-Izzo F. Heavy metal hyperaccumulating plants: How and why do they do it? And what makes them so interesting? Plant Science. 2011;**180**(2):169-181. DOI: 10.1016/j.plantsci.2010.08.016

[5] Koch EW, Barbier EB, Silliman BR, Reed DJ, Perillo GME, Hacker SD, et al. Non-linearity in ecosystem services: Temporal and spatial variability in coastal protection front. Ecology and Environment. 2009;**7**:29-37

[6] Peters EC, Gassman NJ, Firman JC, Richmond RH, Power EA. Ecotoxicology of tropical marine ecosystems. Environmental Toxicology and Chemistry. 1997;**16**:12-40

[7] Blasco F, Saenger P, Janodet E. Mangroves as indicators of coastal change. Catena. 1996;**27**:67-178. DOI: 10.1016/0341-8162(96)00013-6

[8] Kovacs JM, Wang J, Flores-Verdugo F. Mapping mangrove leaf area index at the species level using IKONOS and LAI-2000 sensors for the Agua Brava Lagoon, Mexican Pacific. Estuarine, Coastal, and Shelf Science. 2005;**62**:377-384

[9] Saenger P. Mangrove Ecology, Silviculture, and Conservation. Dordrecht: Klumer Academic Publishers; 2002

[10] Tam NFY, Wong YS. Spatial variation of heavy metals in surface sediments of Hong Kong mangrove swamps. Environmental Pollution. 2000;**110**:195e205

[11] Gonzalez-Mendoza D, Ceja-Moreno V, Gold-Bouchot G, Medrano E-G, Del-Rio M, Valdes-Lozano D, et al. The influence of radical architecture on cadmium bioaccumulation in the black mangrove, *Avicennia germinans* L. Chemosphere. 2007;**67**:330-334

[12] Pourrut B et al. Lead uptake, toxicity, and detoxification in plants. Reviews of Environmental Contamination and Toxicology. 2011;**213**:113-136

[13] Stefaniuk M, Oleszczuk P, Ok YS. Review on nano zerovalent iron (nZVI): From synthesis to environmental applications. Chemical Engineering Journal. 2015;**287**:618-632 DOI: 10.1016/j.cej

[14] Niluxsshun MCD, Masilamani K, Mathiventhan U. Green synthesis of silver nanoparticles from the extracts of fruit pel of *Citrus tangerina*, *Citrus sinensis*, and *Citrus limon* for antibacterial activities. Bioinorganic Chemistry and Applications. 2021;**2021**:6695734. DOI: 10.1155/2021/6695734

[15] Karpagavinayagam P, Vedhi C. Green synthesis of iron oxide nanoparticles using Avicennia marina flower extract. Vacuum. 2018;**2018**:160. DOI: 10.1016/j.vacuum.2018.11.043

[16] Le NT, Dang T-D, Binh KH, Nguyen TM, Xuan TN, La DD, et al.

Green synthesis of highly stable zero-valent iron nanoparticles for organic dye treatment using Cleistocalyx operculatus leaf extract. Sustainable Chemistry, and Pharmacy. 2022;**25**:2352-5541. DOI: 10.1016/j.scp.2022.100598

[17] Hamzezadeh A, Fazlzadeh M, Rahmani K, Poureshgh Y. A novel green synthesis of zero-valent iron nanoparticles (nZVI) using walnut green skin: Characterization, catalytic degradation, and toxicity studies. International Journal of Environmental Analytical Chemistry. 2021. DOI: 10.1080/03067319.2021.1957463

[18] Sravanthi K, Ayodhya D, Yadgiri Swamy P. Green synthesis, characterization of biomaterial-supported zero-valent iron nanoparticles for contaminated water treatment. Journal of Analytical Science and Technology. 2018;**9**:3. DOI: 10.1186/s40543-017-0134-9

[19] Soto-Hidalgo KT, Cabrera CR. Nanoscale Zero Valent Iron for Environmental Cadmium Metal Treatment. InTech. 2018. DOI: 10.5772/intechopen.72737

[20] Soto Hidalgo KT, Guzmán-Blas R, Ortiz-Quiles EO, Fachini ER, Corchado-García J, Larios E, et al. Highly organized nanofiber formation from zero valent iron nanoparticles after cadmium water remediation. RSC Advances. 2015;**5**(4):2777-2784. DOI: 10.1039/c4ra13267j

[21] Sabreena HS, Bhat SA, Kumar V, Ganai BA, Ameen F. Phytoremediation of heavy metals: An indispensable contrivance in green remediation technology. RSC Advances;**11**(9):1255. DOI: 10.3390/plants11091255

[22] Doria Serrano Ma. del Carmen. Química verde: un nuevo enfoque para el cuidado del medio ambiente. Educación química. 2009;**20**(4):412-420. Recuperado en 22 de julio de 2023, de http://www.scielo.org.mx/scielo.php?script=sci_arttext&pid=S0187-893X2009000400004&lng=es&tlng=

Section 3

Strategy A: Safety and Security Assurance

Chapter 5

Green Synthesis of Nanoparticles: A Biological Approach

Rafael Álvarez-Chimal and Jesús Ángel Arenas-Alatorre

Abstract

Nanoparticles are often associated with their small size and numerous applications. However, the synthesis process is equally important as it determines the size and properties of the nanoparticles. While traditional nanoparticle synthesis methods require the use of hazardous chemicals and high-energy consumption, green synthesis offers a sustainable, cost-effective, and environmentally friendly alternative. This approach utilizes natural resources and biologically active compounds that can act as reducing, stabilizing, or capping agents in the one-step synthesis of nanoparticles. Green synthesis offers numerous advantages, including the development of processes with minimal environmental impact and improved safety for nanoparticle synthesis. Overall, the synthesis of nanoparticles using green chemistry is a promising approach for sustainable and efficient production. This chapter provides a general overview of nanoparticles, their applications, and green synthesis, and highlights the various biological resources used in these processes and the factors affecting their synthesis.

Keywords: green synthesis, nanoparticles, plant extract, microorganisms, phytochemicals

1. Introduction

Nanoparticles are small particles with sizes ranging from 1 to 100 nanometers. These materials have gained importance and interest in recent years owing to their large number of applications, because the matter at this scale presents a more compact arrangement of atoms and molecules, generating phenomena and acquiring or enhancing mechanical [1], electrical [2], magnetic [3], optical [4], catalytic [5], and antibacterial [6, 7] properties that are completely different from those of their macroscopic counterparts [8]. They can be classified based on their composition, shape, and size. The most common types of nanoparticles are metals, metal oxides, carbon-based, and quantum dots. Owing to their unique sizes and properties, nanoparticles have attracted significant attention in various fields including medicine, electronics, energy, and environmental science [9, 10]. By reducing their size, nanoparticles can have a higher surface-to-volume ratio, enabling a greater number of atoms or molecules per volume, which means that less material is needed to obtain the same activity and exhibit other properties (**Figure 1**) [11].

Nanoparticles have many potential benefits for the environment. For example, nanoparticles can be used to improve the efficiency of water treatment, air filtration,

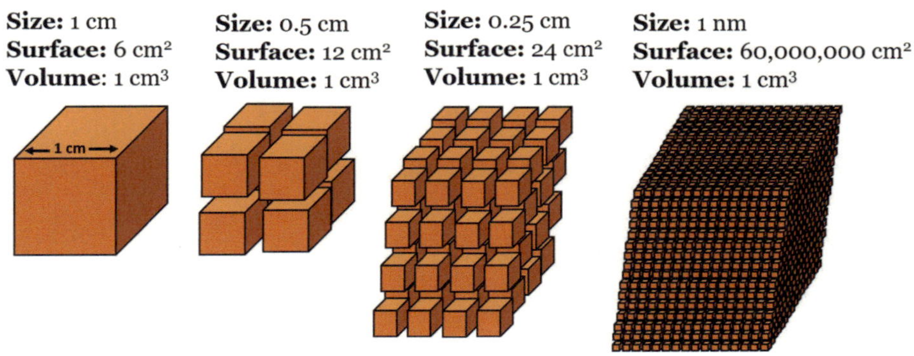

Figure 1.
Surface-to-volume ratio of nanoparticles compared with that of bulk materials.

and soil remediation; reduce pollution, and develop new types of renewable energy technologies [12]. In medicine, nanoparticles have shown potential for drug delivery, imaging, and cancer therapy. They can be functionalized with targeting moieties, making them capable of selectively targeting cancer cells, while sparing normal cells. Additionally, nanoparticles can enhance the efficacy of chemotherapy by improving drug delivery to the tumor site and reducing systemic toxicity [13]. In electronics, nanoparticles are used to fabricate high-performance devices such as sensors, transistors, and solar cells [14]. Nanoparticles have potential applications in fuel cells, hydrogen storage, and catalysis [15].

However, it is also important to address the environmental impact of the nanoparticles. Some studies have shown that nanoparticles can harm plants, animals, and humans, but it depends on many factors, such as concentration, size, and time of exposure [16, 17]. Nanoparticles can easily be released into the environment through various sources, such as industrial emissions, consumer products, and medical procedures. Once released into the environment, nanoparticles can be difficult to control and monitor. There is potential for long-term accumulation. Nanoparticles can accumulate in the environment, and they may be able to persist for long periods. This raises concerns about the potential for nanoparticles to cause long-term harm to the environment and human health [17, 18]. However, one of the alternatives for reducing their environmental impact is to control the synthesis process.

There are many methods for synthesizing nanoparticles, including physical, chemical, and biological processes [19]. Green synthesis, which refers to the eco-friendly and sustainable production of nanoparticles without the use of hazardous chemicals or toxic solvents, has gained attention in recent years within biological processes. Natural sources, such as plants and microorganisms, are popular green synthesis approaches [20]. This method has several advantages over traditional synthesis methods, including low cost, scalability, and reduction of hazardous waste. Moreover, green synthesis can produce nanoparticles with unique shapes, sizes, and surface properties tailored for specific applications [21]. The biological sources used for the green synthesis of nanoparticles contain biologically active compounds, such as enzymes, proteins, polyphenols, flavonoids, and terpenoids, which can act as catalyzing, reducing, stabilizing, or capping agents for one-step synthesis [20, 21].

In summary, this chapter provides a general overview of nanoparticles, their properties and applications, and how green synthesis is used to synthesize them.

This chapter also discusses the different biological resources used for green synthesis, the factors that participate, and the mechanisms involved in their production.

2. Traditional nanoparticle synthesis methods

There are many methods for synthesizing nanoparticles (**Figure 2**). Some of the most common methods include the following.

- Chemical reduction: This method involves the reduction of metal ions in solution using chemical reagents such as sodium borohydride or sodium hydroxide to form nanoparticles [22].
- Coprecipitation: Synthesis involves mixing two or more solutions containing metal ions. When the solutions are mixed, metal ions precipitate out of the solution and form nanoparticles [23].
- Sol-gel: The process requires mixing a metal salt with a solvent and gelling agent. The solvent is evaporated leaving behind the gel. The gel is then heated, causing it to solidify and form nanoparticles [24].
- Microemulsion: This method needs surfactants, water-soluble compounds, and oil-soluble compounds. The mixture forms small droplets that contain the metal ions. When droplets are heated, metal ions precipitate out of the solution and form nanoparticles [25].

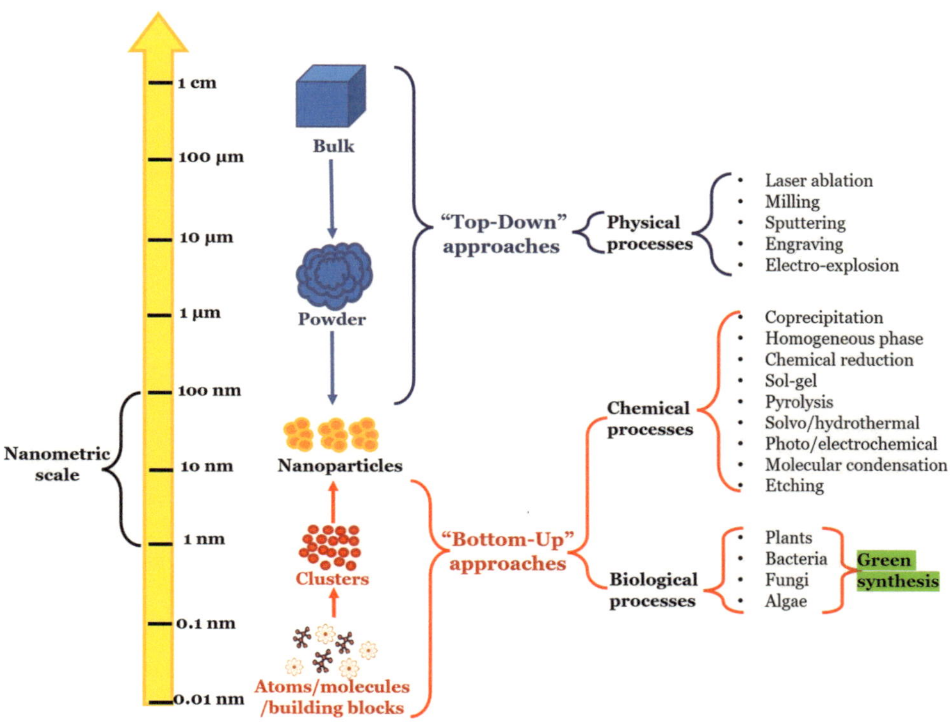

Figure 2.
Nanometric scale and different approaches to nanoparticle synthesis.

- Solvothermal/hydrothermal synthesis: This reaction involves heating a solution of metal ions in water or an organic solvent under high pressure. High pressure and temperature cause metal ions to precipitate out of the solution and form nanoparticles [26].
- Sonochemical/electrochemical synthesis: This process uses ultrasound or an electrical current to break down metal salts into nanoparticles [27].

In addition, there are physical processes, such as laser ablation, milling, and sputtering, where the material is reduced to nanoparticles by the mechanical action of the equipment used [28].

The choice of method depends on the type of nanoparticles being synthesized, the desired size and shape, and the availability of equipment and reagents.

2.1 Environmental limitations in nanoparticle synthesis

Traditional methods for synthesizing nanoparticles have several limitations.

Using organic reagents can harm the environment, humans, and animals, causing illnesses, such as liver damage [18]. In addition, wastewater generated from nanoparticle synthesis can contain harmful chemicals [29].

The low yield is another disadvantage: only a small percentage of the starting materials is converted into nanoparticles, generating raw material waste. The high cost of the starting materials, equipment, labor required, long-time synthesis, and the inability to control the size and shape can limit their applications [30, 31].

2.2 Strategies to overcome barriers to nanoparticle synthesis

Several strategies can be used to overcome the disadvantages of nanoparticle synthesis, such as the use of environmentally friendly solvents, reagents, and processes. Using water, ionic liquids, and supercritical fluids are examples of eco-friendly solvents [21, 32] or we can even perform solvent-free synthesis, eliminating the need for hazardous chemicals and reducing the environmental impact of nanoparticle synthesis [33].

Many nanoparticle synthesis methods are not scalable, which limits their application. Therefore, it is necessary to develop cost-effective and efficient processes to obtain large quantities of nanoparticles [8].

Multipurpose nanoparticles can be used to improve their performance in a variety of applications and fields. For example, biocompatible nanoparticles are used in biomedicine or as stable nanoparticles for long-term applications [34].

The characterization of nanoparticles is important for understanding their size, shape, surface properties, and chemical composition. This information can be used to understand how nanoparticles interact with their environment and ensure they are safe [35].

Strategies to overcome these barriers in nanoparticle synthesis are still under study to develop more innovative, efficient, cost-effective, and environmentally friendly methods.

3. Green synthesis of nanoparticles: an overview

Green synthesis aims to promote innovative chemical technologies to reduce or eliminate the use and production of hazardous substances in the design, manufacture, and use of chemical products. This involves minimizing or, if possible, eliminating the pollution produced in the synthesis processes, avoiding the consumption and wastage of nonrenewable raw materials, using hazardous or polluting materials in product manufacturing, and reducing the synthesis time. Paul J. Anastas, considered the father of green chemistry, defined it as "a work philosophy that involves the use of alternative tools and pathways to prevent pollution," referring to both the design of the synthetic strategy and the treatment of possible secondary products originating from that route [36, 37].

Two approaches can be used to generate nanoparticles [37, 38] (**Figure 2**).

"Top-down" approach: In which nanoparticles are produced using physical techniques such as grinding or abrasion of a material.

"Bottom-up" approach: Where nanoparticles are generated from "building blocks" of atoms or molecules, resulting in more complex assemblies. Three alternatives are identified using this approach.

- Chemical synthesis: The method of producing molecules or particles by the reaction of substances used as raw materials.
- Self-assembly: A technique in which atoms or molecules self-order through physical and/or chemical interactions.
- Positional assembly: The atoms, molecules, and aggregates are deliberately manipulated and positioned individually. However, this method is extremely laborious and unsuitable for industrial applications.

The "bottom-up" approach is preferred over the "top-down" approach because specialized equipment is not required and the time to obtain nanoparticles is shorter. Green synthesis is gaining relevance in producing nanoparticles within the "bottom-up" approach [37].

The use of plant species, algae, or microorganisms such as bacteria or fungi is one of the most commonly used resources for this procedure. Various compounds from plants or microorganisms, including terpenes, polyphenols, alkaloids, carbohydrates, proteins, and genetic materials, play an important role in the synthesis of nanoparticles by acting together [39, 40].

In addition to the biological resources used to perform the synthesis (plants, algae, or microorganisms), other factors influence the shape and size of nanoparticles, such as the concentration of the metal ion, pH, reaction time, and temperature [39, 41].

In general, the phases for the green synthesis of nanoparticles include (**Figure 3**) [39].

- Initial phase: Obtaining the reaction medium, which is the aqueous extract of one or several parts of the plant species or the culture media for the growth of microorganisms, in addition to the precursor salt, which is the source of metal ions.
- Activation phase: Chemical reduction of metal ions and generation of nucleation centers occur where nanoparticles emerge and grow.
- Growth phase: Small adjacent nanoparticles spontaneously fuse into larger particles, forming aggregates, which are influenced by factors such as temperature, concentration, and type of compounds, pH, and reaction time.

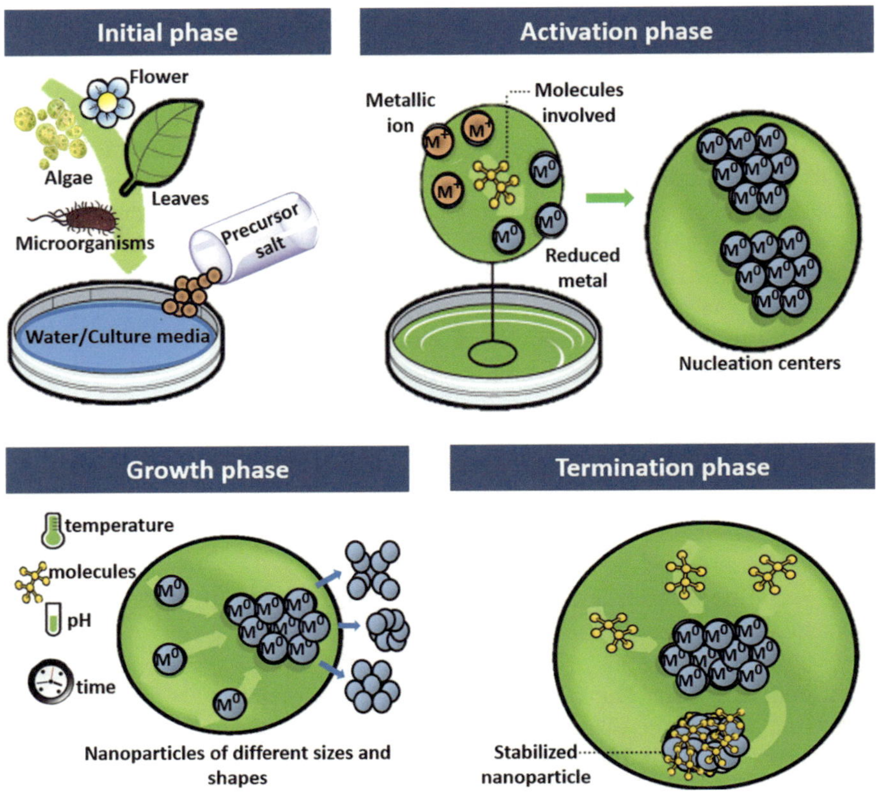

Figure 3.
Phases involved in the green synthesis of nanoparticles.

- Termination phase: The final shape of the nanoparticles is determined, and the compounds that participate in the reaction help stabilize and enhance their properties.

3.1 Biological resources for the green synthesis of nanoparticles

As stated previously, nanoparticles have attracted attention in the fields of biology, medicine, and electronics in recent years, owing to their remarkable applications (**Figure 4**). Numerous nanoparticle synthesis techniques have been developed; however, these may involve the use of toxic compounds and high-energy physical processes. An alternative is the use of biological methods to circumvent these obstacles. Bacteria, fungi, algae, and plant species are some of the most commonly used biological resources for the green synthesis of nanoparticles (**Figure 4**). This biological approach has provided a method that is reliable, straightforward, benign, and environmentally beneficial [40, 42].

3.1.1 Bacteria

Nanoparticle synthesis using bacteria is performed both extracellularly and intracellularly [38].

Intracellular: The synthesis is carried out inside the living microorganism, using its growth conditions to favor synthesis, known as "nanoparticle micro-factories." To recover nanoparticles, bacteria must be destroyed [43].

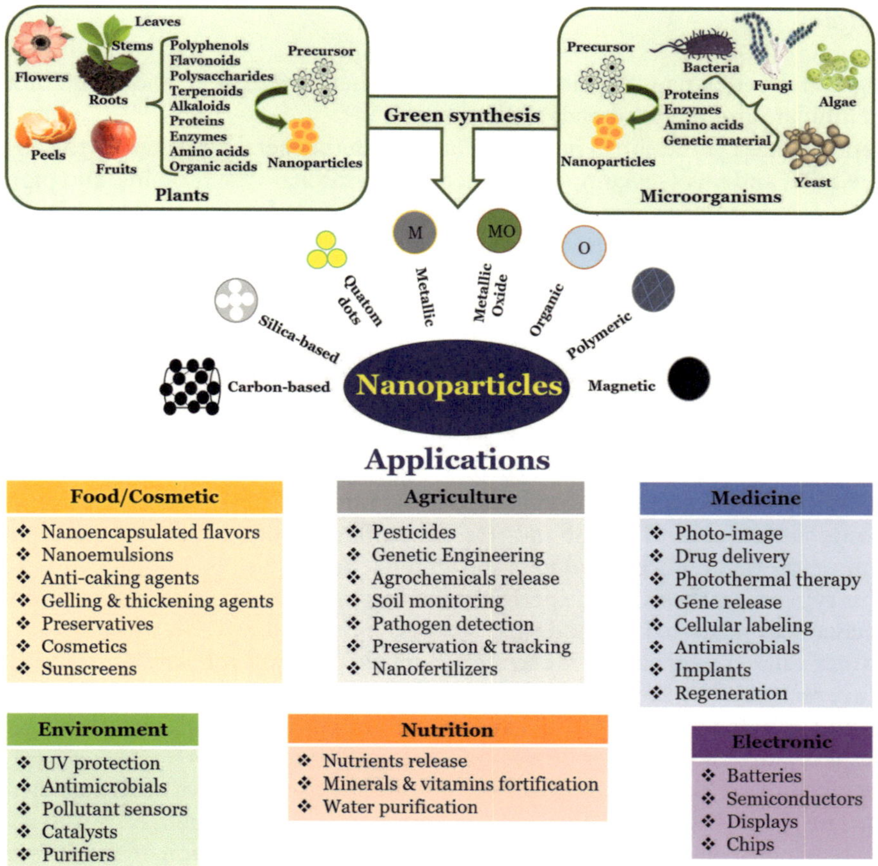

Figure 4.
Biological resources and compounds used for the green synthesis of nanoparticles and some of their applications [9].

Extracellular: The components released by the bacteria after lysis are used. The synthesis is performed by adding a metal salt precursor to the medium in which these components are located. Extracellular synthesis has the advantage of being faster because it does not require additional steps to recover nanoparticles from microorganisms [43, 44].

Enzymes, such as reductases, which catalyze the reduction of metal ions into nanoparticles, participate in the synthesis. Even components of the genetic material participate in this process [45, 46].

3.1.2 Fungi

Fungi contain active biomolecules, such as proteins or enzymes, that participate in nanoparticle synthesis, improving their yields and stability [47].

Some fungal species can synthesize nanoparticles using extracellular amino acids. For example, glutamic and aspartic acids on the surface of yeast or the reductase enzyme in the cytosol of fungi reduce metal ions to form nanoparticles. This is facilitated by the presence of hydroxyl groups in the mycelium, which donate electrons to the metal ion and reduce it to form nanoparticles. Aliphatic and aromatic amines or some proteins act as coating agents to stabilize them [48, 49].

3.1.3 Algae

Algae are used in nanotechnology because of their low toxicity and their ability to bioaccumulate and reduce metals [50].

Nanoparticle synthesis can be intracellular, with the metal ion entering the alga, or extracellular, and involves compounds such as polysaccharides, proteins, and pigments that direct the reduction of metal ions and coat the newly formed nanoparticles. These particles are subsequently released from the cell in the form of colloids [51].

3.1.4 Plant species

The use of plants in nanoparticle synthesis is one of the most widely used methods because of its environmentally friendly nature, as it avoids the use of toxic or harmful substances. It is also one of the fastest and most economical methods because it involves fewer steps [39, 40]. This makes it highly efficient in the nanoparticle production process compared to synthesis using microorganisms.

Plants contain several compounds (terpenes, flavonoids, polyphenols, alkaloids, proteins, etc.) that reduce metal ions and stabilize the resulting nanoparticles [52].

This type of synthesis can be performed using intracellular, extracellular, and phytochemical-mediated methods [53].

Intracellular: The synthesis is carried out inside the plant cell, and the nanoparticles are recovered by breaking down the structure, which is very similar to the intracellular method using microorganisms. Control of the growth factors of plant species is required so that they do not interfere with synthesis [54].

Extracellular: This method is the most commonly used because of its ease and speed. The process begins by obtaining a plant extract, which is generally water-based, to which a metal salt precursor is added. Owing to the action of the different compounds present in the extract, nanoparticles are generated and stabilized in a single step [54, 55].

Phytochemically mediated: This is based on the extracellular method, but with the difference that isolated phytochemical compounds are used and other substances are added to stabilize the nanoparticles. There is greater control over the synthesis, but more components and steps are involved [53].

3.2 Factors involved in the green synthesis of nanoparticles

As in any synthesis process, reaction conditions, such as temperature, pH, and reaction time, play an important role in the shape, size, and yield of the synthesized nanoparticles [39–41] (**Figure 3**).

Temperature: This is one of the most influential factors, as the shape (spherical, prismatic, flakes, triangular, octahedral, etc.), size, and synthesis depend on temperature. As the temperature increases, the reaction rate and the formation of nucleation centers increase, resulting in higher yields. Different temperatures promote different interactions between the reactants, giving rise to various shapes; the larger the temperature increase, the larger the size of the nanoparticles [56, 57].

pH: This influences the nucleation centers, generating more centers at higher pH values. Another important influence of pH is that some nanoparticles can only be synthesized in acidic or alkaline media. For example, magnetic nanoparticles are synthesized at an alkaline pH, and metal oxide nanoparticles are generally synthesized at an acidic or neutral pH [58].

Time: This parameter plays an important role in defining the size of the nanoparticles. It has been observed that longer reaction times favor an increase in the size of the nanoparticles and higher yields, owing to the prolonged interaction time between reactants [59].

3.3 The mechanism involved in the green synthesis of nanoparticles

The plant extract or organism used for the synthesis is an important factor that influences the morphology and size of nanoparticles because different concentrations of metabolites or cellular components give rise to differences in the nanoparticles [40, 60] (**Figure 5**).

Proteins and enzymes facilitate the formation of nanoparticles from metal ions. Because of their high reducing activity, proteins and enzymes can attract metal ions to specific regions of a molecule responsible for reduction, facilitating the formation of nanoparticles; however, their chelating activity is not excessive. The amino acids of a protein can greatly influence the size, morphology, and quantity of nanoparticles generated, thus playing a very important role in determining their shape and yield.

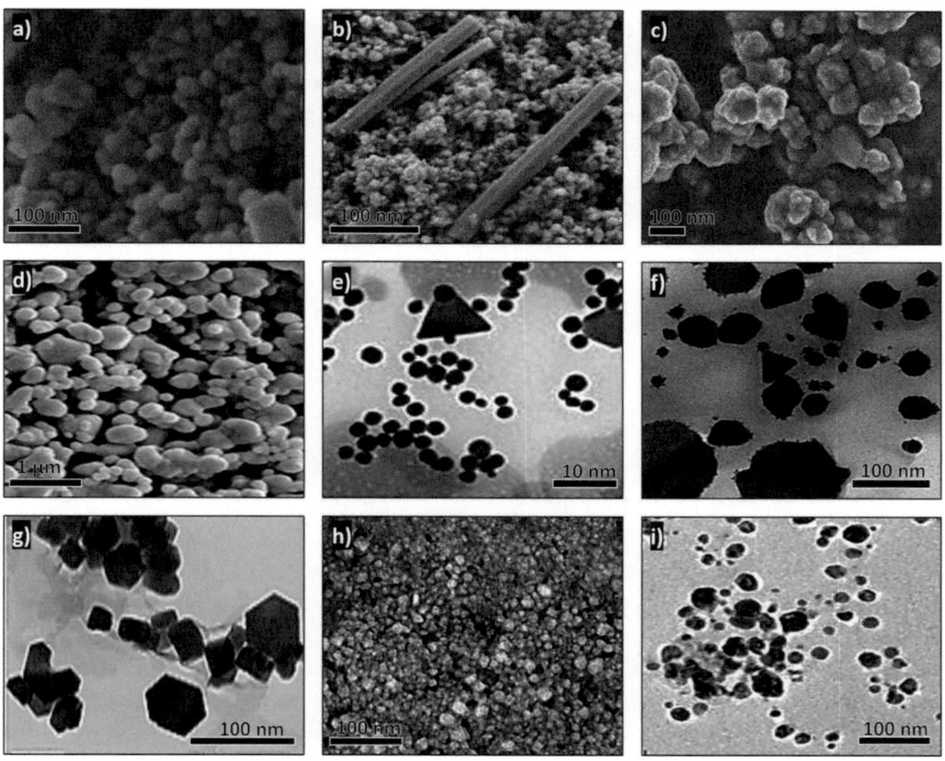

Figure 5.
Green-synthesized nanoparticles. (a) Spherical ZnO nanoparticles using the leaves of Dysphania ambrosioides (plant). (b) Prismatic ZnO nanoparticles using the stems and leaves of Dysphania ambrosioides (plant). (c) Quasi-spherical Fe_3O_4 nanoparticles using the leaves of Datura innoxia (plant). (d) Quasi-spherical Ag nanoparticles using stems of Aloe vera (plant) [61]. (e) Spherical and triangular Au nanoparticles using Lentinula edodes (fungus) [43]. (f) Irregular Ag and triangular Au nanoparticles using Ganoderma lucidum (fungus) [43]. (g) Hexagonal MgO nanoparticles using the flowers of Saussurea costus (plant) [62]. (h) Irregular Cu nanoparticles using Salmonella typhimurium (bacterium) [63]. (i) Quasi-spherical Ag nanoparticles using Dunaliella salina (alga) [64].

Removing a proton from amino acids or other molecules results in the formation of resonant structures capable of further oxidation. This process is accompanied by the active reduction of metal ions followed by the formation of nanoparticles [39].

Flavonoids are a large group of polyphenolic compounds that can actively chelate and reduce metal ions because they contain multiple functional groups capable of forming these structures. Structural transformations of flavonoids also generate protons that reduce metal ions to form nanoparticles; therefore, they are involved in the nucleation stage, their formation, and further aggregation. Saccharides can also play a role in nanoparticle formation. Monosaccharides, such as glucose, can act as reducing agents, as the aldehyde group of the sugar is oxidized to a carboxyl group through the addition of hydroxyl groups, which in turn leads to the reduction of metal ions and the synthesis of nanoparticles [39].

The mechanism of green synthesis of nanoparticles has been associated with the action of polyphenols, which act as ligands. Metal ions form coordination compounds, in which the fundamental structural unit is the central metal ion surrounded by coordinated groups arranged spatially at the corners of a regular tetrahedron. The aromatic hydroxyl groups in polyphenols bind to metal ions and form stable coordinated complexes. This system undergoes direct decomposition at high temperatures, releasing nanoparticles from the complex system [65].

Flavonoids, amino acids, proteins, terpenoids, tannins, and reducing sugars have hydroxyl groups that surround the metal ions to form complexes. After this process, the hydroxyl ions are oxidized to carbonyl groups, which stabilize the nanoparticles. Synthesis is favored if the participating molecules have at least two hydroxyl groups at the ortho- and para-positions [52, 65].

Amino acids influence the size, morphology, and yield of nanoparticles generated [23], depending on the specific amino acids present in the extract and their concentration, along with the reaction conditions that give rise to nanoparticles with different shapes [65].

4. Confirming that the biological approach of nanoparticle synthesis is a green chemistry method

To corroborate that the processes of nanoparticle synthesis using biological resources are "green synthesis methods," the 12 principles mentioned above are revisited [66–68] (**Table 1**).

Considering the above, the 12 principles of green synthesis are fulfilled using biological resources, such as plants, bacteria, fungi, and algae, to synthesize nanoparticles [69–71].

Finally, green synthesis of nanoparticles is a sustainable and environmentally friendly alternative to traditional methods of nanoparticle synthesis. Traditional methods often take long periods of time, use toxic chemicals and solvents, or generate waste products that can pollute the environment and pose health risks to humans and animals. In contrast, the green synthesis method uses renewable natural resources, such as plant extracts and microorganisms, which are less damaging and can be replenished over time. In addition, these methods are often more cost-effective and faster than traditional procedures because they do not require expensive chemicals or equipment and are considered one-step syntheses, which contribute to energy savings [72].

I. The plant extract or culture medium used in the synthesis can be easily disposed of, either by using it for composting in the case of plant extract or by sterilizing the culture medium and similarly disposed of without causing environmental harm.	**II.** The synthesis yields are high, which supports the great incorporation of the raw material into the final product.
III. Few or no toxic wastes are generated because aqueous extracts or culture media are used and treated after use.	**IV.** Nanoparticles have the same or better properties than those generated using other methods.
V. Water is preferably used as the solvent.	**VI.** Syntheses are carried out at room temperature; although the temperature is a factor that influences synthesis, depending on the nanoparticles to be synthesized, it may be necessary to vary the synthesis temperature.
VII. Vegetal extracts and microbial culture media can be reused more than once for several syntheses.	**VIII.** The method focuses on nanoparticle synthesis; there are no subproducts, or those generated are the metabolites that participate in the reaction and can be reused in the synthesis.
IX. Synthesis is catalyzed by biological compounds found in organisms. Making a faster or one-step synthesis.	**X.** Nanoparticles should be handled with the necessary precautions, regardless of the process used in their synthesis.
XI. There is considerable control over the synthesis process; it can be stopped at any time if there is a problem and resumed without issue. The reactants and raw materials are handled with the necessary care, and the generated products are easily treated.	**XII.** The risk of accidents is minimized because the synthesis is performed at ambient temperature and pressure.

Table 1.
The 12 principles of green synthesis are fulfilled with the biological approach to produce nanoparticles.

In furtherance of these advantages, green synthesis methods are still being developed to improve their efficiency and scalability, leading to the potential benefits of green synthesis of nanoparticles or even their application to the synthesis of other molecules as drugs or nutraceuticals.

5. Conclusion

Nanoparticles have emerged as a versatile and promising class of materials with unique properties that can be harnessed for various applications. The use of green synthesis utilizing natural resources and biologically active compounds to produce nanoparticles is an area of continuous research to improve processes, reduce environmental damage, and meet the increasing demand for the application of these nanostructures. Utilizing biological resources, the synthesis of nanoparticles is inexpensive, faster, and considered a one-step synthesis while preserving or even improving the physical and chemical properties of the nanoparticles. With the great potential of this method and the sustainable and efficient production of nanoparticles, different sizes and shapes can be obtained, which makes it a very attractive option not only for the synthesis of nanostructures, but also for the application of this technique in the synthesis of other compounds.

Acknowledgements

The authors acknowledge Dr. Samuel Tehuacanero Cuapa, Physicist. Roberto Hernández Reyes, and Arq. Diego Quiterio Vargas for their technical support.

Thanks to the Consejo Nacional de Humanidades, Ciencias y Tecnologías (CONAHCYT) for the scholarship granted to Rafael Álvarez-Chimal with the CVU number: 579637.

Funding was provided by the *UNAM-DGAPA*-PAPIIT project IN112422.

Conflict of interest

The authors declare no conflicts of interest.

Author details

Rafael Álvarez-Chimal* and Jesús Ángel Arenas-Alatorre
Laboratory 113 Synthesis of Magnetic Nanomaterials, Condensed Matter Department, Institute of Physic, National Autonomous University of Mexico, Ciudad Universitaria, Mexico City, Coyoacán, Mexico

*Address all correspondence to: ac.rafael.11@gmail.com

References

[1] Wu Q, Shou MW, Du ZY, Jun GH, Hui D. Mechanical properties of nanomaterials: A review. Nanotechnology Reviews. 2020;**9**(1): 259-273. DOI: 10.1515/ntrev-2020-0021

[2] Min SH, Lee TH, Lee S, Song JH, Lee GY, Zontar D, et al. Simulation of electrical conductivity for nanoparticles and nanotubes composite sensor according to geometrical properties of nanomaterials. Composites Part B: Engineering. 2019;**174**:107003. DOI: 10.1016/j.compositesb.2019.107003

[3] Martínez-Mera I, Espinosa-Pesqueira ME, Pérez-Hernández R, Arenas-Alatorre J. Synthesis of magnetite (Fe_3O_4) nanoparticles without surfactants at room temperature. Materials Letters. 2007;**61**(23-24):4447-4451. DOI: 10.1016/j.matlet.2007.02.018

[4] Kelly KL, Coronado E, Zhao LL, Schatz GC. The optical properties of metal nanoparticles: The influence of size, shape, and dielectric environment. The Journal of Physical Chemistry. B. 2003;**107**(3):668-677. DOI: 10.1021/jp026731y

[5] Zeng J, Zhang Q, Chen J, Xia Y. A comparison study of the catalytic properties of Au-based nanocages, nanoboxes, and nanoparticles. Nano Letters. 2010;**10**(1):30-35. DOI: 10.1021/nl903062e

[6] Álvarez-Chimal R, García-Pérez VI, Álvarez-Pérez MA, Arenas-Alatorre JÁ. Green synthesis of ZnO nanoparticles using a *Dysphania ambrosioides* extract. Structural characterization and antibacterial properties. Materials Science and Engineering: C. 2021;**118**:111540. DOI: 10.1016/j.msec.2020.111540

[7] Reyes-Carmona L, Camps E, Campos-González E, Mercado-Celis G, Cervantes-Garduño A, Pérez-Ibarra EA, et al. Antimicrobial evaluation of bismuth subsalicylate nanoparticles synthesized by laser ablation against clinical oral microorganisms. Optics and Laser Technology. 2023;**158**:108930. DOI: 10.1016/j.optlastec.2022.108930

[8] Baig N, Kammakakam I, Falath W. Nanomaterials: A review of synthesis methods, properties, recent progress, and challenges. Materials Advances. 2021;**2**(6):1821-1871. DOI: 10.1039/D0MA00807A

[9] Chaudhry N, Dwivedi S, Chaudhry V, Singh A, Saquib Q, Azam A, et al. Bio-inspired nanomaterials in agriculture and food: Current status, foreseen applications and challenges. Microbial Pathogenesis. 2018;**123**:196-200. DOI: 10.1016/j.micpath.2018.07.013

[10] Kolahalam LA, Kasi Viswanath IV, Diwakar BS, Govindh B, Reddy V, Murthy YLN. Review on nanomaterials: Synthesis and applications. Materials Today: Proceedings. 2019;**18**:2182-2190. DOI: 10.1016/j.matpr.2019.07.371

[11] Khan I, Saeed K, Khan I. Nanoparticles: Properties, applications and toxicities. Arabian Journal of Chemistry. 2019;**12**(7):908-931. DOI: 10.1016/j.arabjc.2017.05.011

[12] Lu AH, Salabas EL, Schüth F. Magnetic nanoparticles: Synthesis, protection, functionalization, and application. Angewandte Chemie, International Edition. 2007;**46**(8):1222-1244. DOI: 10.1002/anie.200602866

[13] Zhang L, Gu F, Chan J, Wang A, Langer R, Farokhzad O. Nanoparticles in

medicine: Therapeutic applications and developments. Clinical Pharmacology and Therapeutics. 2008;**83**(5):761-769. DOI: 10.1038/sj.clpt.6100400

[14] Gao J, Gu H, Xu B. Multifunctional magnetic nanoparticles: Design, synthesis, and biomedical applications. Accounts of Chemical Research. 2009;**42**(8):1097-1107. DOI:. 10.1021/ar9000026

[15] Singh R, Altaee A, Gautam S. Nanomaterials in the advancement of hydrogen energy storage. Heliyon. 2020;**6**(7):e04487. DOI: 10.1016/j.heliyon.2020.e04487

[16] Yao D, Chen Z, Zhao K, Yang Q, Zhang W. Limitation and challenge faced to the researches on environmental risk of nanotechnology. Procedia Environmental Sciences. 2013;**18**:149-156. DOI: 10.1016/j.proenv.2013.04.020

[17] Wilson N. Nanoparticles: Environmental problems or problem solvers? Bioscience. 2018;**68**(4):241-246. DOI: 10.1093/biosci/biy015

[18] Ray PC, Yu H, Fu PP. Toxicity and environmental risks of nanomaterials: Challenges and future needs. Journal of Environmental Science and Health, Part C. 2009;**27**(1):1-35. DOI: 10.108010590500802708267/

[19] Iravani S, Korbekandi H, Mirmohammadi SV, Zolfaghari B. Synthesis of silver nanoparticles: Chemical, physical and biological methods. Research in Pharmaceutical Sciences. 2014;**9**(6):385-406

[20] Mustapha T, Misni N, Ithnin NR, Daskum AM, Unyah NZ. A review on plants and microorganisms mediated synthesis of silver nanoparticles, role of plants metabolites and applications. International Journal of Environmental Research and Public Health. 2022;**19**(2):674. DOI: 10.3390/ijerph19020674

[21] Shafey AME. Green synthesis of metal and metal oxide nanoparticles from plant leaf extracts and their applications: A review. Green Processing and Synthesis. 2020;**9**(1):304-339. DOI: 10.1515/gps-2020-0031

[22] Manikam VR, Cheong KY, Razak KA. Chemical reduction methods for synthesizing Ag and Al nanoparticles and their respective nanoalloys. Materials Science and Engineering B. 2011;**176**(3):187-203. DOI: 10.1016/j.mseb.2010.11.006

[23] Nam NH, Luong NH. Nanoparticles: Synthesis and applications. In: Materials for Biomedical Engineering. Amsterdam, Netherlands: Elsevier; 2019. pp. 211-240. DOI: 10.1016/B978-0-08-102814-8.00008-1

[24] Bokov D, Turki Jalil A, Chupradit S, Suksatan W, Javed Ansari M, Shewael IH, et al. Nanomaterial by sol-gel method: Synthesis and application. Advances in Materials Science and Engineering. 2021;**2021**:1-21. DOI: 10.1155/2021/5102014

[25] Cid A. Synthesis of NPs by microemulsion method. In: Microemulsion - A Chemical Nanoreactor. London, UK: IntechOpen; 2018. DOI: 10.5772/intechopen.80633

[26] Ndlwana L, Raleie N, Dimpe KM, Ogutu HF, Oseghe EO, Motsa MM, et al. Sustainable hydrothermal and solvothermal synthesis of advanced carbon materials in multidimensional applications: A review. Materials. 2021;**14**(17):5094. DOI: 10.3390/ma14175094

[27] Ashassi-Sorkhabi H, Rezaei-moghadam B, Bagheri R, Abdoli L,

Asghari E. Synthesis of Au nanoparticles by thermal, sonochemical and electrochemical methods: Optimization and characterization. Polymerase Chain Reaction. 2015;**3**(1):24-34. DOI: 10.22036/pcr.2015.7311

[28] Krishnia L, Thakur P, Thakur A. Synthesis of Nanoparticles by Physical Route. Synthesis and Applications of Nanoparticles. Singapore: Springer Nature Singapore; 2022. pp. 45-59. DOI: 10.1007/978-981-16-6819-7_3

[29] Khoshnamvand M, Hao Z, Fadare OO, Hanachi P, Chen Y, Liu J. Toxicity of biosynthesized silver nanoparticles to aquatic organisms of different trophic levels. Chemosphere. 2020;**258**:127346. DOI: 10.1016/j.chemosphere.2020.127346

[30] Rahimi HR, Doostmohammadi M. Nanoparticle synthesis, applications, and toxicity. In: Applications of Nanobiotechnology. London, UK: IntechOpen; 2020. DOI: 10.5772/intechopen.87973

[31] Jamkhande PG, Ghule NW, Bamer AH, Kalaskar MG. Metal nanoparticles synthesis: An overview on methods of preparation, advantages and disadvantages, and applications. Journal of Drug Delivery Science and Technology. 2019;**53**:101174. DOI: 10.1016/j.jddst.2019.101174

[32] Ying S, Guan Z, Ofoegbu PC, Clubb P, Rico C, He F, et al. Green synthesis of nanoparticles: Current developments and limitations. Environmental Technology and Innovation. 2022;**26**:102336. DOI: 10.1016/j.eti.2022.102336

[33] Landge S, Ghosh D, Aiken K. Solvent-Free Synthesis of Nanoparticles, Green Chemistry. Amsterdam, Netherlands: Elsevier; 2018. pp. 609-646. DOI: 10.1016/B978-0-12-809270-5.00022-4

[34] Kim D, Shin K, Kwon SG, Hyeon T. Synthesis and biomedical applications of multifunctional nanoparticles. Advanced Materials. 2018;**30**(49):1802309. DOI: 10.1002/adma.201802309

[35] Halamoda-Kenzaoui B, Vandebriel RJ, Howarth A, Siccardi M, David CAW, Liptrott NJ, et al. Methodological needs in the quality and safety characterisation of nanotechnology-based health products: Priorities for method development and standardisation. Journal of Controlled Release. 2021;**336**:192-206. DOI: 10.1016/j.jconrel.2021.06.016

[36] Anastas PT, Warner JC. Green Chemistry: Theory and Practice. Oxford [England], New York: Oxford University Press; 1998. p. 135

[37] Singh J, Dutta T, Kim KH, Rawat M, Samddar P, Kumar P. 'Green' synthesis of metals and their oxide nanoparticles: Applications for environmental remediation. Journal of Nanbiotechnology. 2018;**16**(1):84. DOI: 10.1186/s12951-018-0408-4

[38] Singh A, Gautam PK, Verma A, Singh V, Shivapriya PM, Shivalkar S, et al. Green synthesis of metallic nanoparticles as effective alternatives to treat antibiotics resistant bacterial infections: A review. Biotechnology Reports. 2020;**25**:e00427. DOI: 10.1016/j.btre.2020.e00427

[39] Makarov VV, Love AJ, Sinitsyna OV, Makarova SS, Yaminsky IV, Taliansky ME, et al. "Green" nanotechnologies: Synthesis of metal nanoparticles using plants. Acta Naturae. 2014;**6**(1):35-44. DOI: 10.32607/20758251-2014-6-1-35-44

[40] Hebbalalu D, Lalley J, Nadagouda MN, Varma RS. Greener techniques for

the synthesis of silver nanoparticles using plant extracts, enzymes, bacteria, biodegradable polymers, and microwaves. ACS Sustainable Chemistry & Engineering. 2013;**1**(7):703-712. DOI: 10.1021/sc4000362

[41] Agarwal H, Venkat Kumar S, Rajeshkumar S. A review on green synthesis of zinc oxide nanoparticles – An eco-friendly approach. Resource-Efficient Technologies. 2017;**3**(4):406-413. DOI: 10.1016/j.reffit.2017.03.002

[42] Velusamy P, Kumar GV, Jeyanthi V, Das J, Pachaiappan R. Bio-inspired green nanoparticles: Synthesis, mechanism, and antibacterial application. Toxicological Research. 2016;**32**(2):95-102. DOI: 10.5487/TR.2016.32.2.095

[43] Vetchinkina E, Loshchinina E, Kupryashina M, Burov A, Pylaev T, Nikitina V. Green synthesis of nanoparticles with extracellular and intracellular extracts of basidiomycetes. PeerJ. 2018;**6**:e5237. DOI: 10.7717/peerj.5237

[44] Das VL, Thomas R, Varghese RT, Soniya EV, Mathew J, Radhakrishnan EK. Extracellular synthesis of silver nanoparticles by the *Bacillus* strain CS 11 isolated from industrialized area. 3 Biotechnology. 2014;**4**(2):121-126. DOI: 10.1007/s13205-013-0130-8

[45] Singh P, Kim YJ, Zhang D, Yang DC. Biological synthesis of nanoparticles from plants and microorganisms. Trends in Biotechnology. 2016;**34**(7):588-599. DOI: 10.1016/j.tibtech.2016.02.006

[46] Messaoudi O, Bendahou M. Biological synthesis of nanoparticles using endophytic microorganisms: Current development. In: Nanotechnology and the Environment. London, UK: IntechOpen; 2020. DOI: 10.5772/intechopen.93734

[47] Mukherjee P, Ahmad A, Mandal D, Senapati S, Sainkar SR, Khan MI, et al. Fungus-mediated synthesis of silver nanoparticles and their immobilization in the mycelial matrix: A novel biological approach to nanoparticle synthesis. Nano Letters. 2001;**1**(10):515-519. DOI: 10.1021/nl0155274

[48] Syed A, Ahmad A. Extracellular biosynthesis of platinum nanoparticles using the fungus *fusarium oxysporum*. Colloids and Surfaces. B, Biointerfaces. 2012;**97**:27-31. DOI: 10.1016/j.colsurfb.2012.03.026

[49] Riddin TL, Gericke M, Whiteley CG. Analysis of the inter- and extracellular formation of platinum nanoparticles by *fusarium oxysporum* f. sp. *lycopersici* using response surface methodology. Nanotechnology. 2006;**17**(14):3482-3489. DOI: 10.1088/0957-4484/17/14/021

[50] Rana A, Yadav K, Jagadevan S. A comprehensive review on green synthesis of nature-inspired metal nanoparticles: Mechanism, application and toxicity. Journal of Cleaner Production. 2020;**272**:122880. DOI: 10.1016/j.jclepro.2020.122880

[51] Dahoumane SA, Yéprémian C, Djédiat C, Couté A, Fiévet F, Coradin T, et al. A global approach of the mechanism involved in the biosynthesis of gold colloids using micro-algae. Journal of Nanoparticle Research. 2014;**16**(10):2607. DOI: 10.1007/s11051-014-2607-8

[52] Carrillo-López LM, Soto-Hernández RM, Zavaleta-Mancera HA, Vilchis-Néstor AR. Study of the performance of the organic extracts of *Chenopodium ambrosioides* for Ag nanoparticle synthesis. Journal of Nanomaterials. 2016;**2016**:1-13. DOI: 10.1155/2016/4714162

[53] Dauthal P, Mukhopadhyay M. Noble metal nanoparticles: Plant-mediated

synthesis, mechanistic aspects of synthesis, and applications. Industrial and Engineering Chemistry Research. 2016;**55**(36):9557-9577. DOI: 10.1021/acs.iecr.6b00861

[54] Saim AK, Kumah FN, Oppong MN. Extracellular and intracellular synthesis of gold and silver nanoparticles by living plants: A review. Nanotechnology for Environmental Engineering. 2021;**6**(1):1. DOI: 10.1007/s41204-020-00095-9

[55] Naikoo GA, Mustaqeem M, Hassan IU, Awan T, Arshad F, Salim H, et al. Bioinspired and green synthesis of nanoparticles from plant extracts with antiviral and antimicrobial properties: A critical review. Journal of Saudi Chemical Society. 2021;**25**(9):101304. DOI: 10.1016/j.jscs.2021.101304

[56] Álvarez-Chimal R, García-Pérez VI, Álvarez-Pérez MA, Tavera-Hernández R, Reyes-Carmona L, Martínez-Hernández M, et al. Influence of the particle size on the antibacterial activity of green synthesized zinc oxide nanoparticles using *Dysphania ambrosioides* extract, supported by molecular docking analysis. Arabian Journal of Chemistry. 2022;**15**(6):103804. DOI: 10.1016/j.arabjc.2022.103804

[57] Thanh NTK, Maclean N, Mahiddine S. Mechanisms of nucleation and growth of nanoparticles in solution. Chemical Reviews. 2014;**114**(15):7610-7630. DOI: 10.1021/cr400544s

[58] Handayani W, Ningrum AS, Imawan C. The role of pH in synthesis silver nanoparticles using *Pometia pinnata* (Matoa) leaves extract as bioreductor. Journal of Physics: Conference Series. 2020;**1428**(1):012021. DOI: 10.1088/1742-6596/1428/1/012021

[59] De Oliveira RC, Amoresi RAC, Marana NL, Zaghete MA, Ponce M, Chiquito AJ, et al. Influence of synthesis time on the morphology and properties of CeO_2 nanoparticles: An experimental–Theoretical study. Crystal Growth and Design. 2020;**20**(8):5031-5042. DOI: 10.1021/acs.cgd.0c00165

[60] Kuppusamy P, Yusoff MM, Maniam GP, Govindan N. Biosynthesis of metallic nanoparticles using plant derivatives and their new avenues in pharmacological applications – An updated report. Saudi Pharmaceutical Journal. 2016;**24**(4):473-484. DOI: 10.1016/j.jsps.2014.11.013. DOI: 10.1016/j.molstruc.2016.12.069

[61] Tippayawat P, Phromviyo N, Boueroy P, Chompoosor A. Green synthesis of silver nanoparticles in *aloe vera* plant extract prepared by a hydrothermal method and their synergistic antibacterial activity. PeerJ. 2016;**4**. DOI: 10.7717/peerj.2589

[62] Amina M, Al Musayeib NM, Alarfaj NA, El-Tohamy MF, Oraby HF, Al Hamoud GA, et al. Biogenic green synthesis of MgO nanoparticles using *Saussurea costus* biomasses for a comprehensive detection of their antimicrobial, cytotoxicity against MCF-7 breast cancer cells and photocatalysis potentials. PLoS One. 2020;**15**(8). DOI: 10.1371/journal.pone.0237567

[63] Ghorbani HR. Extracellular synthesis of copper nanoparticles using culture supernatants of *salmonella typhimurium*. Oriental Journal of Chemistry. 2015;**31**(1):527-529. DOI: 10.13005/ojc/310165

[64] Singh AK, Tiwari R, Kumar V, Singh P, Riyazat Khadim SK, Tiwari A, et al. Photo-induced biosynthesis of silver nanoparticles from aqueous extract of *Dunaliella salina* and their anticancer potential. Journal of Photochemistry

and Photobiology, B: Biology. 2017;**166**:202-211. DOI: 10.1016/j.jphotobiol.2016.11.020

[65] Nava OJ, Luque PA, Gómez-Gutiérrez CM, Vilchis-Nestor AR, Castro-Beltrán A, Mota-González ML, et al. Influence of *Camellia sinensis* extract on zinc oxide nanoparticle green synthesis. Journal of Molecular Structure. 2017;**1134**:121-125

[66] Duan H, Wang D, Li Y. Green chemistry for nanoparticle synthesis. Chemical Society Reviews. 2015;**44**(16):5778-5792. DOI: 10.1039/C4CS00363B

[67] Razavi M, Salahinejad E, Fahmy M, Yazdimamaghani M, Vashaee D, Tayebi L. Green Chemical and Biological Synthesis of Nanoparticles and their Biomedical Applications. Green Processes for Nanotechnology. Cham: Springer International Publishing; 2015. pp. 207-235. DOI: 10.1007/978-3-319-15461-9_7

[68] Badria F, AbouHabieb M, Bar FA. Synthesis of Nanoparticles Using Green Chemistry Green Synthesis. 1. Auflage ed. München: Grin Verlag; 2019

[69] Chopra H, Bibi S, Singh I, Hasan MM, Khan MS, Yousafi Q, et al. Green metallic nanoparticles: Biosynthesis to applications. Frontiers in Bioengineering and Biotechnology. 2022;**10**:874742. DOI: 10.3389/fbioe.2022.874742

[70] Michael A, Singh A, Roy A, Islam MR. Fungal- and algal-derived synthesis of various nanoparticles and their applications. Bioinorganic Chemistry and Applications. 2022;**2022**:1-14. DOI: 10.1155/2022/3142674

[71] Raj S, Trivedi R, Soni V. Biogenic synthesis of silver nanoparticles, characterization and their applications—A review. Surfaces. 2021;**5**(1):67-90. DOI: 10.3390/surfaces5010003

[72] Kebede MA, Wubieneh TA, Yohannes YB, Shah KJ. Green synthesis of zinc oxide from aqueous fruit extract of *Dovyalis abyssinica* (Koshem) and application for water purification. Ethiopian Journal of Science and Technology. 2023;**16**(1):1-12. DOI: 10.4314/ejst.v16i1.1

Chapter 6

Green Synthesis of Copper, Silver, and Gold Nanoparticles: Applications in Medicinal Chemistry

Faisal Nawaz, Irum Jamil and Bushra Shakoor

Abstract

For the creation of an effective "green chemistry," scientists have paid a lot of attention to a method for creating metal nanoparticles in recent years. The topic of nanoscale metal synthesis is one that is now pertinent due to the growing use of nanoscale metals in industries like engineering and medicine and the environment. The development of nanotechnology has revolutionized society and has focused on a number of scientific fields, including cancer, human epidemiology, and material science. Utilizing biological agents to prepare metallic nanoparticles through green chemistry reduces negative effects and improves the metal's anticancer properties. Commercially, nanoparticles are typically prepared *via* physical, chemical, and biological techniques. For metal and metal oxide nanoparticles made from natural extracts, such as gold, silver, and copper oxide, which are useful to improve biomedical applications like antibacterial, antifungal, and antioxidant, have been summarized in this chapter. The easy availability of plants that are safe, easy to handle, and inexpensive allows for the synthesis of many nanoparticles. These techniques are environment-friendly, nontoxic, and economically viable green synthesized processes.

Keywords: green synthesis, nanoparticles, antifungal activity, antibacterial activity, antioxidant activity

1. Introduction of nanoparticles

Nanotechnology has advanced to a level that has never been seen before in recent decades. Therefore, it is not surprising that researchers are becoming more interested in the subject. Nanotechnology is the branch of science and technology concerned with the design, creation, and application of nanoparticles [1]. In general, items that are considered to be nanoscale or nanometer in size should fall within the range of 1–1000 nm, where "nano" is defined as 10^9. However, according to the recommendations of the European Commission, a nanomaterial is any substance, whether organic, inorganic, or synthetic, that contains at least 50% of particles (in unbound or aggregated state) with one or more exterior dimensions in the range of 1–100 nm [2].

Due to a specific behavior under light irradiation, namely the formation of localized surface plasmon resonance (LSPR), which gives these materials their peculiar optical properties, the nanoparticles formed from coinage metals draw attention.

Of all the "noble" metals, silver has the highest thermal and electrical conductivities as well as the lowest melting and boiling temperatures. The "noble" metals with the highest reactivity, silver also has harmful effects on a variety of microbes thanks to its cations [3]. Solar energy harvesting and electronic gadgets are just two examples of the many technological and medical applications that silver in the form of nanoparticles unveiled.

Functionalized nanoparticles with various targeting moieties have a wide range of uses in biomedicine, including diagnosis, targeting, drug/nucleic acid administration, imaging, and therapy. Furthermore, NPs are used as sensitive probes in Raman scattering and imaging applications using the surface-enhanced Raman scattering approach [4]. The application of NPs in photothermal therapy, radiation therapy, computed tomography, biosensors, and other domains has greatly boosted their potential in biomedical fields [5] as shown in **Figure 1**. NPs-based biosensors with excellent sensitivity and selectivity are being developed due to their intrinsic electrical and optical properties, as well as their ability to conjugate with various biomolecules. Over the last decade, NPs-based biosensors have received a lot of interest in the diagnosis of numerous diseases. Recently, several NPs-based biological assays for the detection and quantification of analytes in urinary samples have been highlighted, with an emphasis on protein analysis. Such NP-based assays are beneficial in the diagnosis of a variety of ailments, including renal problems, cancer, and heart disease [6].

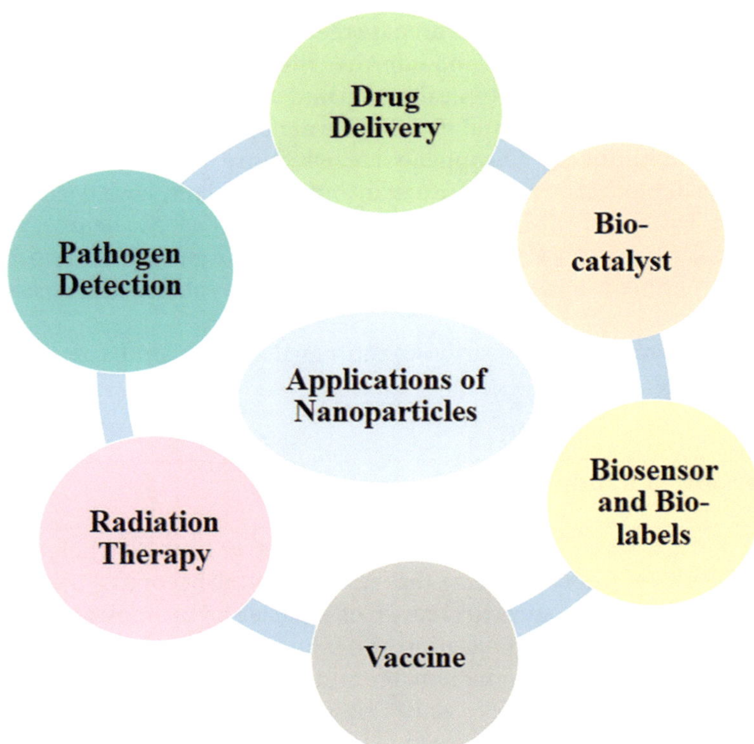

Figure 1.
Some important applications of nanoparticles in medicinal chemistry.

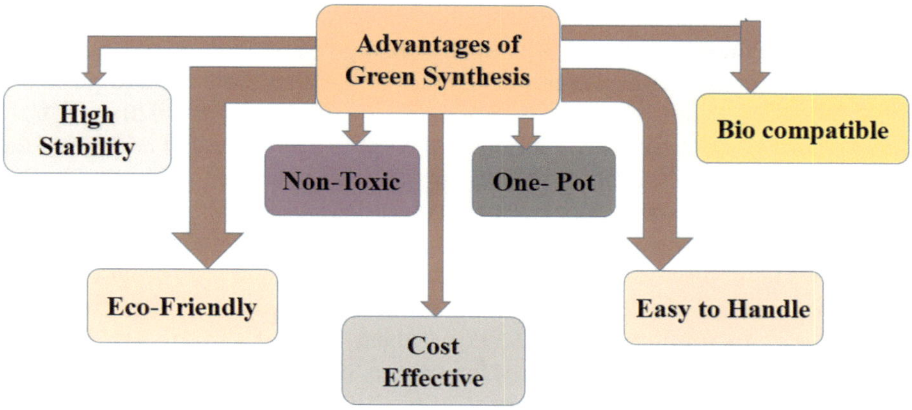

Figure 2.
Advantages of green synthesis.

It is now widely accepted that nanotechnologies have the potential to be useful in a number of industries, including pharmaceutical research, water purification, and information and communication technologies and the development of stronger and lighter materials. Nanotechnologies produce and manipulate materials at the nanoscale by scaling up from small groups of atoms or by reducing or refining bulk materials [7].

The concept of "green chemistry" for "sustainable development" has received a lot of attention in the recent decade. Sustainable development is defined as development that fulfills current demands while taking into account future generations' abilities to meet their own [8]. Green chemistry is a growing area that advocates for the implementation of concepts aimed at minimizing the usage and manufacture of harmful chemical compounds [9]. As a result, greener methods reduce the environmental impact of industrial activities. Researchers created these technologies to propose potential solutions to the costly processes and harmful compounds discovered when using classic synthesis methods [10]. Green nanomaterial synthesis is the optimal way for lowering the risk level of nanotechnology by minimizing the negative repercussions of its creation and use. **Figure 2** highlight the primary advantages of green chemistry approaches.

In this chapter, authors have discussed the various biological properties of transition metals and their oxides. We have also explained their synthesis methods and different applications in medicinal chemistry like, antibacterial, antifungal, and antioxidant to stops the different chain reactions occurring in biological systems.

2. Properties of metal NPs

2.1 Silver nanoparticles

It has been marketed that silver nanoparticles (AgNPs) are a superior antibacterial agent that can fight infections-causing bacteria both *in vitro* and *in vivo*. The antibacterial properties of AgNPs extend to both Gram-negative and Gram-positive bacteria, including strains that are multidrug resistant. AgNPs have a number of simultaneous mechanisms of action, and they work synergistically against harmful bacteria like

Escherichia coli and *Staphylococcus aureus* when paired with antibiotics or other antibacterial agents like chemical compounds. AgNPs are highly suited for usage in medical and healthcare goods because of their characteristics, which allow them to successfully treat or prevent infections. A new generation of powerful antibacterial drugs [11].

The exploration of silver nanoparticles (AgNPs) can be found in various products within the medical and healthcare sectors, including surgical and food handling tools, clothing, cosmetics, dental products, catheters, and dressings. The improved antibacterial activity of Ag at the nanoscale has proven to be most beneficial [12].

Numerous plant diseases pose a persistent danger to agricultural production. Despite a large increase in pesticide application, disease losses continue to be a major problem. Losses are predicted to be 21.5% for wheat, 22.5% for maize, and 30% for rice globally [12]. Fungi, which account for around 80% of plant infections, are the predominant group of pathogens in cereal crops [13]. According to recent estimates, there are 6.2 million different species of fungi, demonstrating how common they are [14]; because it works well against 650 different bacteria, silver can be employed in the plant protection industry [15].

2.2 Gold nanoparticles

Gold nanoparticles (AuNPs) are renowned for having outstanding physical and chemical properties. The optical properties of AuNPs are among the most researched physical characteristics in the biological field. Interestingly, while the color of gold in its bulk form is yellow, the color of finely separated AuNPs can vary according on their size, from violet to wine red. These color fluctuations in AuNPs can be linked to surface plasmon resonance (SPR) and or LSPR, two photophysical processes. Surface plasmon polaritons, sometimes referred to as plasmon waves, travel across a thin metal surface and cause SPR. Plasmon oscillation in a metal nanocrystal is known as LSPR. Applications for effective biological sensing can be made with both LSPR and SPR. It is believed that LSPR is responsible for the color shift from bright yellow (in bulk) to various size-dependent dazzling colors. A plasmonic phenomenon results from the shrinkage of particle size below the incident electromagnetic radiation's wavelength [4].

AuNPs as a whole are nontoxic; however, the stabilizing capping agents and chemical byproducts employed to make AuNPs can make the AuNP solution poisonous. For instance, even at a nanomolar concentration, cetyl trimethyl ammonium bromide (CTAB), which is used to stabilize gold nanorods, is hazardous. However, toxicity can be decreased by switching out a harmful capping agent for a suitable biocompatible one or by altering CTAB to stop it from dissolving. It was discovered that citrate-coated AuNPs are noncytotoxic at low concentrations in a paper on the cytotoxicity of these particles. The size and concentration of AuNPs both affect toxicity. From the standpoint of their biomedical engineering applications, it is crucial to investigate how AuNPs interact with the components of biological medium. It has been discovered that AuNPs with a size of less than 2 nm destroy biological cells by oxidatively harming the mitochondrial structure.

2.3 Copper nanoparticles

Copper is one of the substances that is the most common on the planet. It has had a huge impact on history due to its many qualities, which include excellent electrical and thermal conductivity, remarkable corrosion resistance, and improved malleability. It has been used as jewelry, swords, and currency from the early fourteenth century [16].

Property	Silver nanoparticle	Gold nanoparticle	Copper nanoparticle
Color	Grayish-white to black [25]	Reddish to yellow [26]	Reddish-brown [27]
Melting point	961.8°C [25]	1064.18°C [26]	1984°C [27]
Conductivity	Excellent [25]	Excellent [26]	Good [27]
Reactivity	Moderate [25]	Low [26]	High [27]
Stability	Susceptible to oxidation [28]	Highly stable [26]	Susceptible to oxidation [29]
Applications	Antimicrobial, electronics [28]	Biomedical, catalysis [26]	Electronics, catalysis [29]
Optical properties	Strong plasmonic effects [28]	Strong plasmonic effects [26]	Plasmonic effects [29]

Table 1.
Some general properties of silver, gold, and copper nanoparticles.

Copper, which is necessary for the metabolism of living things, is present in more than 30 different types of proteins. Numerous enzymes with copper as a component support a number of physiological functions, including oxygen transport and iron homeostasis [17]. In addition, copper is present in numerous human organs, bones, and skin [18]. When taken in quantities that are too great for the body to handle, copper turns poisonous and can cause hemolysis, jaundice, abdominal pain, nausea, and, in extreme circumstances, death [19]. Tap water is a typical source of copper poisoning because the pipe used for water delivery either is made of copper alloys or contains copper [20]. Contrarily, human copper insufficiency is a fairly rare disorder.

Numerous studies in both humans and animals have demonstrated that a copper deficit can result in issues with the connective tissues, osteoporosis and other bone diseases, as well as a higher risk of infection [21]. Anemia and poor fetal development can result from copper shortage, according to a different study [22]. Copper nanoparticles (CuNPs) have attracted interest from the general public because of their capacities in the mechanical, electrical, magnetic, and thermal realms. They have also been used in water treatment, surgical tool coatings, and heat transfer systems to stop the growth of microbes [23]. Using copper has the advantage of being inexpensive and widely accessible, which lowers the cost of purchasing CuNPs. One of the disadvantages of CuNPs is their susceptibility to oxidation when exposed to water environments. Copper transforms into CuO and Cu_2O and transforms into Cu^{2+} during preparation, making it challenging to continue synthesizing CuNPs in an ambient environment [24]. Some general properties of silver, gold, and copper nanoparticles have been presented in **Table 1**.

3. Synthesis of nanoprticles

3.1 Synthesis of AgNPs by traditional route

3.1.1 By chemical method

Silver nitrate solution (0.001 M) and sodium borohydride (0.1 M) were utilized as a metal salt precursor and a reducing agent, respectively, for the chemical production of AgNPs. Stabilizing agent polyethylene glycol 80 was utilized at a concentration of 50 mM [30].

3.2 Synthesis of AgNPs by green route

3.2.1 By phytochemical method

By reducing a silver nitrate solution at 0.001 M in the presence of the *Lysiloma* acapulcensis extract, AgNPs were prepared. 2.5 mL of *Lysiloma* acapulcensis aqueous extract was mixed with 2.5 mL of $AgNO_3$ in a 1:1 ratio for 2 min at room temperature. The reaction solution was then made with distilled water to a final volume of 10 mL, and the solutions were exposed to white light for 15, 30, and 60 min. UV–Vis spectroscopy with a wavelength range of 300–600 and a resolution of 1 nm was used to track the development of the synthesis. A blank was created using the aqueous extract. There were three duplicates of each experiment. With a resolution of 400–4000 cm^{-1}, FT-IR spectroscopy analysis was used to identify the functional group of biomolecules present in the *Lysiloma acapulcensis* aqueous extract at room temperature [30].

3.2.2 By using microorganisms

In this mechanism of synthesis, bacteria break down Ag^+ ions to their elemental form (Ag^0), which builds up outside the cell and produces AgNPs. Depending on the culture medium used to cultivate the bacteria, the extracellularly generated AgNPs might be hexagonal, spherical, triangular, circular, disc-shaped, or cuboidal. The reducing agent for the biogenic reduction of Ag^+ to Ag^0 is either a tiny soluble secreted enzyme or a protein found on the bacterial cell wall. Outside of the cell, *Bacillus licheniformis*, *Bacillus pumilus*, and *Bacillus persicus* all produce AgNPs that range in size from 72 to 92 nm [31]. AgNPs produced during extracellular production can be easily recovered *via* high-speed centrifugation (10,000 to 12,000 rpm), where they are retained in a pellet that can be redissolved in any favored solvent.

3.2.2.1 Mechanism of AgNP production

Bottom-up strategies used by bacteria to synthesize AgNPs from small building blocks like atoms and molecules using redox reactions have been shown in **Figure 3**. The biogenic reduction of silver ions takes place at or near ambient pressure, temperature,

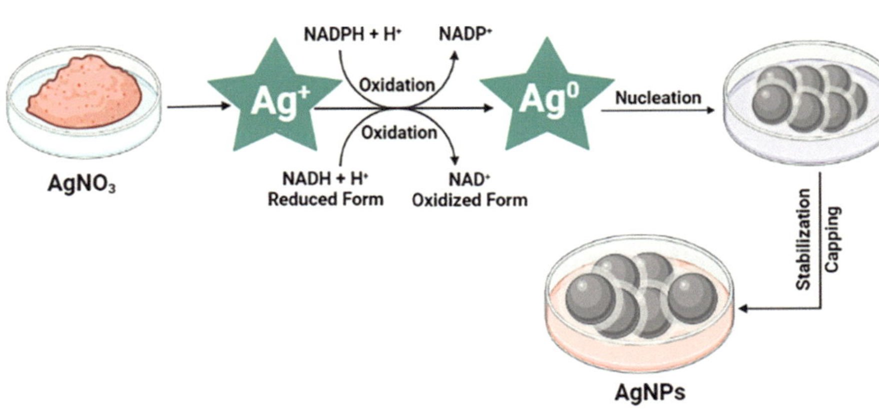

Figure 3.
Mechanism of AgNP production by bacteria mediated via NADH dependent nitrate reductase enzyme created with BioRender.com [32].

and pH. Silver ions are trapped, reduced, and capped, and crystals are stabilized as part of the procedure. According to certain theories, in the event of intracellular synthesis, silver ions are brought into the cell for reduction into elemental silver [32].

Some bacteria produce a transmembrane proton gradient that is actively supported by Na and Ag ions from the extracellular environment. The uptake of silver ions inside the cell and the initiation of AgNP synthesis are caused by a number of silver-binding membrane proteins that draw silver ions and use the energy from adenosine triphosphate hydrolysis to do so. The cell secretes extracellular polymeric materials made of proteins and polysaccharides that can bind to and capture ions. Reducing substances including proteins, enzymes, carbohydrates, and amino acids that are also secreted from the cells work to lower the trapped ions outside the cells. Only a few studies have demonstrated the importance of nicotinamide adenine dinucleotide (NAD) + hydrogen (H) NADH-dependent reductase in the synthesis of AgNPs, where NADH-H^+ contributes its electrons to silver ions, which are then reduced to their elemental form and accumulate as AgNPs [32].

3.3 Synthesis of AuNPs by traditional route

3.3.1 By chemical method

Choline chloride and $ZnCl_2$, which have comparable physical and chemical properties, were combined to create the quaternary ammonium-based room temperature ionic liquids (QAILs). The QAILs were prepared by mixing choline chloride and ZnCl2 in a molar ratio of 1:1 (QAIL-I) and 1:2 (QAIL-II), which have similar physical and chemical properties. The QAILs were subjected to an overnight hoover treatment at 90°C to get rid of moisture and volatile contaminants before use. The liquids can be heated to at least 190°C and are at least as thermally stable as $AlCl_3$-based liquids. Deionized water was used to dissolve $HAuCl_4$ in order to create a 10.0 mmol $l^{-1}Au^{3+}$ solution. A series of QAIL stabilized AuNP colloids were created by adding 0.2 mL of Au^{3+} solution drop-by-drop into 10.0 mL of QAILs at various temperatures, with the Au^{3+} content held constant at 0.2 mmol l^{-1}. These freshly generated AuNP colloids were then immediately

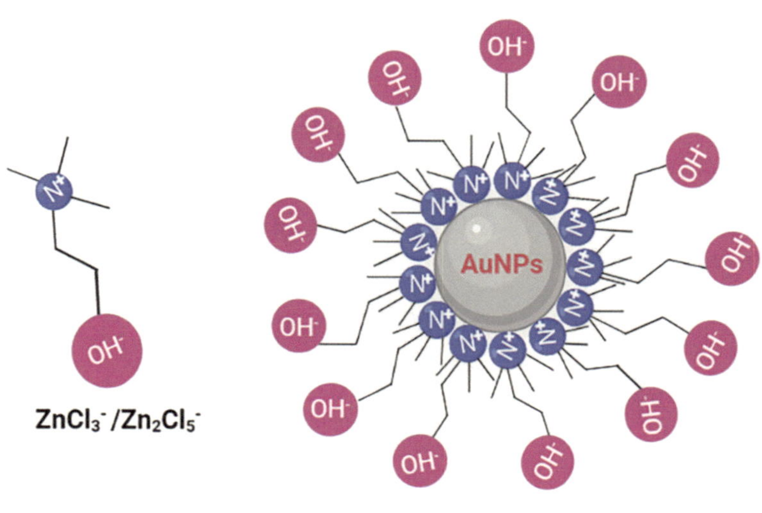

Figure 4.
Molecular structures of QAILs and stabilized AuNPs created with BioRender.com [33].

supplemented with an additional 0.2 mL of Au^{3+} while being kept at the same temperature. For 15 min for each specimen, the QAILs and AuNP colloids were heated in an oil-bath heating reactor while being rapidly agitated in a glass beaker. For additional measurements, the AuNPs were centrifuged to methanol with a dilution ratio of 1:4 after the QAIL-AuNP colloid specimens had been separated with methanol [33].

3.3.1.1 Synthesis mechanism

The two-step process of nucleation and subsequent development of the particles is the most widely acknowledged mechanism for the formation of metal nanoparticles. The metal cores are gradually accumulated by the freshly reduced atoms, and occasionally cluster stabilizers participate in the reduction of some of the metal ions to zero valency metal nanoparticles. An organic cation and an inorganic anion form the structure of the QAIL given in **Figure 4** [33].

3.4 Synthesis of AuNPs by green route

3.4.1 By phytochemical method

The plant leaf (*Magnolia kobus*) broth solution was made by boiling 5 g of fully cleaned, finely chopped leaves in 100 mL of pH 7.0 phosphate buffered saline for 5 min and then decanting the solution. 1 mM $KAuCl_4$ was added to the leaf broth to reduce the Au^{3+} ions. By conducting the reaction in a water bath at 95°C with reflux, the effects of temperature on the synthesis rate and particle size/shape of the produced AuNPs were investigated. By repeatedly centrifuging the AuNPs solution at 15,000 rpm for 20 min and redispersing the pellet in deionized water, the AuNPs solution was purified [34].

3.4.2 By using microorganisms

Pseudomonas aeuroginosa is obtained from the Department of Microbiology and Immunology in Egypt. Burns are the source of two clinical samples of bacterial isolates used in this investigation. The isolates were identified as *Pseudomonas aeruginosa* ATCC 90271 by microbiology and biochemistry and were utilized as a reference strain [35].

Pseudomonas aeruginosa control strains, and the two isolates were used. At 37°C, the bacteria were incubated while growing aerobically in a 50 mL nutritional broth medium. It was shaken for 24 h at 150 rpm. After the incubation, the overnight bacterial culture was centrifuged at 5000 rpm for 5 min to get the supernatants. In order to create AuNPs, 50 mL of cell-free supernatant was combined with hydrogen tetrachloroaurate to produce a final concentration of gold ions of 1 mM. The resulting solution was then incubated at 37°C for 24 h. The cell-free supernatant containing nanoparticles was recovered after 24 h of incubation [35].

3.5 Synthesis of CuNPs by traditional route

3.5.1 By chemical method

Copper(II) sulfate pentahydrate was used as a precursor salt in the chemical reduction procedure to prepare the Cu nanoparticles, and starch was used as a capping agent. The first step in the preparation process is to add 120 mL of starch (1.2%) solution to a 0.1 M copper(II) sulfate pentahydrate solution while vigorously stirring for 30 min. The synthesis solution is mixed rapidly and continuously as 50 mL of a 0.2 M ascorbic acid solution is added in the second stage. The resulting solution was then progressively supplemented with 30 mL of a 1 M sodium hydroxide solution while being heated at 80°C for 2 h. The color of solution was changed from yellow to brown. After the reaction was finished, the solution was removed from the heat and allowed to settle for the night. The next day, the supernatant solution was carefully thrown away. Filtration was used to separate the precipitates from the mixture, and the excess starch linked to the nanoparticles was removed by washing the precipitates three times with deionized water and ethanol. Precipitates of brown color were produced, and they were dried at room temperature. Nanoparticles were stored in glass vials for later examination after drying [36].

3.6 Synthesis of CuNPs by green route

3.6.1 By phytochemical method

The *Kigelia Africana* fruits were properly washed, cut into slices, and sun-dried for 5 days to ensure that all moisture was removed. After that, the fruits were pulverized using a mechanical grinder. In order to perform the extraction, 50 g of the powdered plant fruit were added to a 1000 mL beaker along with 500 mL of ethanol. To achieve thorough extraction, the beaker holding the ethanol and powdered fruit was kept for 48 h while being repeatedly stirred with a magnetic stirrer every day. After filtering, the extract was put into an airtight container for storage [37].

3.6.2 By using microorganisms

Using *pseudomonas fluorescens*, copper sulfate ($CuSO_4$) solution was treated separately with cell pellet and cell-free supernatant to prepare CuNPs. The culture was centrifuged at 7500 rpm for 20 min at 30°C after 48 h, and both the cell pellet and the cell-free culture supernatant were collected and used separately. 1 g of biomass was separately suspended in 20 mL of solutions containing 318, 750, and 1000 ppm ($CuSO_4$). The preparation of CuNP was assessed after supernatant had been collected, centrifuged at 10,000 rpm for 20 min at 30°C. Varied concentrations of $CuSO_4$ solution were mixed with varied volumes of cell-free supernatant, and the mixture was shaken in an incubator for 24–48 h at 30°C and 150 rpm [38].

4. Applications of NPs

4.1 Applications of AgNPs

4.1.1 Mechanism of action against bacteria

The precise cause of AgNPs' antibacterial impact on bacteria is yet unclear. We describe a potential mechanism of AgNP action in **Figure 5** that could lead to antibacterial activity. AgNPs continually release silver ions, which might be a method for destroying microorganisms. Since silver ions are more intimately related to sulfur proteins and are likewise attracted to them electrostatically, they can readily cling

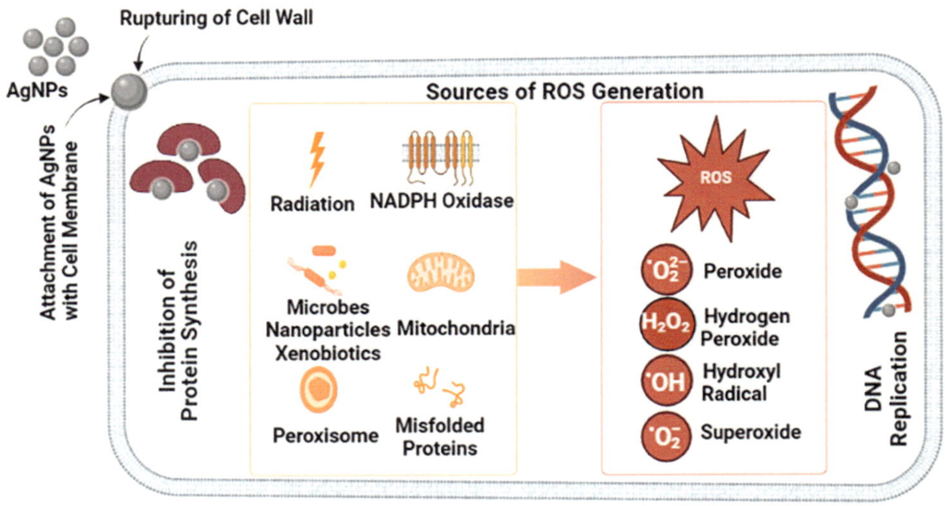

Figure 5.
Action mechanism of AgNPs against the bacteria and the biofilm created with BioRender.com [39].

to the cytoplasmic membrane and cell wall. Silver ions attaching to the cell wall or cytoplasmic membrane boost the permeability of the cell, which ultimately results in cell breakdown, and disrupt the bacterial envelope at the same time. Free silver ions disable respiratory enzymes as they enter cells, producing reactive oxygen species that stop the synthesis of adenosine triphosphate. The primary species that initiates deoxyribonucleic acid (DNA) alteration and cell membrane rupture is reactive oxygen species (ROS). Phosphorus and sulfur are crucial elements in DNA. Even yet, the interaction of AgNPs with the sulfur and phosphorus in DNA can make it difficult to replicate DNA and support new cells, or even cause bacteria to die. Because the silver ions can prevent protein production, the ribosome can occasionally get denaturated in the cytoplasm [39].

4.1.2 Mechanism of action against fungus

The size, shape, and coating agents have a significant impact on the antifungal activity of biogenic AgNPs. Deciphering a single mechanism of action is particularly challenging due to the large diversity of biogenic AgNPs. For this reason, the majority of the present study has concentrated on figuring out how chemically synthesized AgNPs work, which is attributed to the AgNPs adhering to the surface of the fungus due to electrostatic attraction as presented in **Figure 6**. Ag^+ is dynamically released from extracellular AgNP accumulation, enters the cell, and increases the intracellular concentration as well as the intracellular manufacture of AgNPs. No cell receptors or membrane channels for the absorption of silver have been identified to yet. However, it has been discovered that Ag^+ is imported through the high-affinity copper transporter (Ctr1) [40].

4.1.3 Mechanism of action against cancer cells

Apoptosis, or programmed cell death, is avoided by cancerous cells, which keep multiplying. The development of cancer therapies is mostly focused on

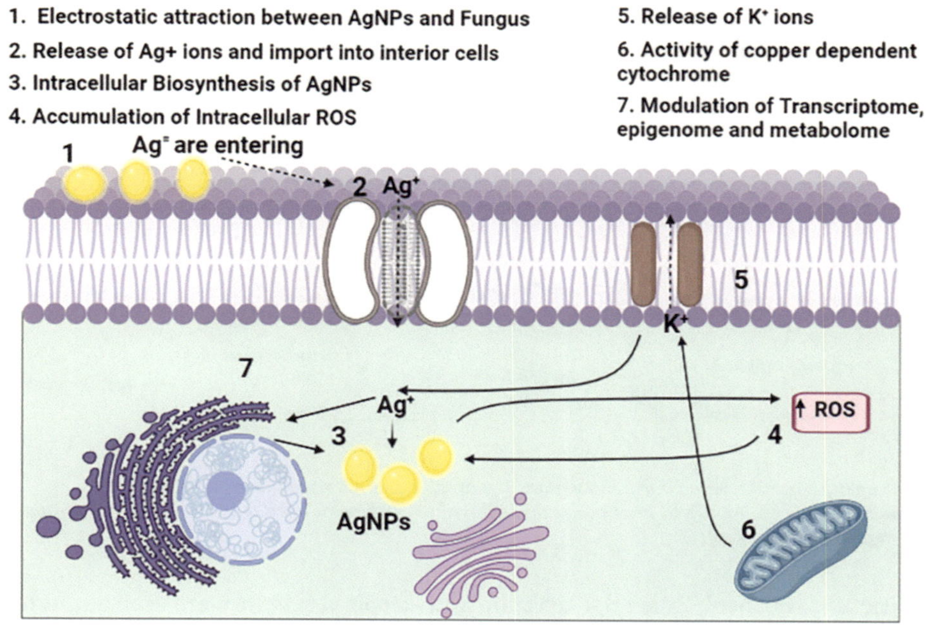

Figure 6.
Action mechanism of AgNPs against fungi [40].

the aforementioned characteristics of cancer cells. To effectively combat cancer, plant-based nanosized silver is gaining popularity. There are two signaling pathways—intrinsic and extrinsic—that can be activated to cause apoptosis. Apoptosis, which is a limiting process in malignant cells, is triggered by DNA damage or extreme cell stress [41].

Beginning with various steps of apoptotic protein activation, DNA damage, mitochondrial disintegration, the development of an apoptosome, and finally cell shrinkage, apoptosis is a natural process in all living things. These turn into the most crucial targets to be used in cancer therapy. When applied to certain target locations, AgNP exhibits anticancer action. According to recent studies, AgNPs primarily affects ROS, upping oxidative stress and DNA synthesis. Normal cellular homeostasis, which is essential for cell survival, is maintained by ROS [42].

4.2 Applications of AuNPs

4.2.1 Mechanism of action against bacteria

In order to prepare effective functionalized metallic nanoparticles, it is important to take into account that Gram-positive and Gram-negative bacteria have different cell wall and membrane structures [43]. The boronic acid group of phenylboronic acid is highly bound by the glycolipid-containing bacterial cell membrane presented in **Figure 7(a)** [44]. As a result of interaction with lipoteichoic acid and lipopolysaccharide of Gram-positive and Gram-negative bacteria, respectively, AuNPs were changed with thiol- and amine-tethered phenylboronic acids. Nanoparticles have been prepared and modified using natural substances derived from bacteria, fungus, and plants [46]. For instance, sildenafil citrate, geniposidic acid, 3,5-dimethylphenol,

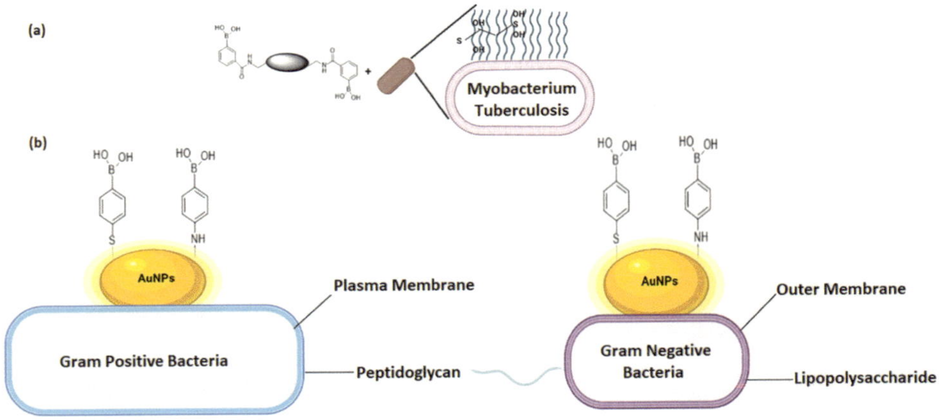

Figure 7.
(a) Multimeric boronic acid to target cell-envelope glycans of Mycobacterium tuberculosis [44] (b) functionalization of the AuNPs by amine- and thioltethered phenylboronic acids to target Gram-positive and Gram-negative bacteria [45].

palmitic acid, borneol, 2-hexyl-1-octanol, and -terpinyl acetate were used as phytochemicals in an aqueous leaf extract from *Acalypha indica* that was used to synthesize spherical AuNPs with a 20 nm size shown in **Figure 7(b)** [45].

4.2.2 Antifungal activity

AuNPs produced from cyanobacteria have been shown to have antifungal action; however, this activity was nanoparticle-size dependent. For instance, AuNPs of size 7 nm were more effective than those of size 15 nm. Candida was more effectively controlled by 25 nm particles than by 30 nm particles in this situation. Additionally, the entire AuNPs test had demonstrated action based on nanoparticle dose or concentration. The primary reason for the antifungal effect of AuNPs was the potential interaction with fungal protein. The metabolically active H^+- adenosine triphosphatase enzyme in fungi is interacted with by AuNPs, which changes the enzyme's usual alignment and causes a loss of fungi activity. This disturbance in metabolic activity prevents fungi from absorbing nutrients, which ultimately results in fungi cell death. Furthermore, the efficient antifungal activity may have been caused by a unique interaction involving shapes and sizes between AuNPs and the components of the plasma membrane, such as sulfur-containing proteins or phosphorus of bases in DNA. AuNPs impair normal processes like as synthesis, replication, and repair, which results in cell death [47].

4.2.3 Antioxidant activity

Recently, AuNPs have been considered a top choice for delivering a range of drugs to their intended locations. These payloads range in size from tiny drug molecules to bigger macromolecules including proteins, ribonucleic acid, and DNA [48]. The oxidative stress brought on by ROS such hydroxyl, epoxyl, peroxylnitrile, superoxide, and singlet oxygen plays a significant role in a variety of pathological disorders, such as inflammatory processes, atherosclerosis, aging, cancer, and neurological illnesses. Nucleic acid damage and enzyme inactivation are caused by excess ROS or oxidative stress that affects the host's antioxidant system [49].

DOI: http://dx.doi.org/10.5772/intechopen.1002859

4.3 Applications of CuONPs

4.3.1 Antibacterail activity

The size of nanoparticles is important in many applications, including medicine delivery. The surface area that may be targeted for medication release increases with decreasing particle size. Smaller nanoparticles have greater mobility and the capability to enter and move about within the bacterial cell compartments, which makes their average size crucial for copper oxide nanoparticles (CuONPs') potential antibacterial action. As a result, they are powerful antibacterial agents. However, various types of anionic compounds, such as organic acids, albumins, surfactants, polymers, and others, can easily prevent electrostatic adhesion from occurring in a solution. This phenomenon influences how different biomolecules, such as proteins and carbohydrates, which can adsorb on nanoparticles and produce a corona with different surface properties from the original nanoparticles, interact with the nanoparticles. This is most likely the cause that in biological fluids, formulations containing anionic polyelectrolytes, and surfactant-containing formulations, CuONPs can rapidly lose their antibacterial copper oxide efficacy [50].

One putative antibacterial mechanism is the release of free Cu^{2+} ions from CuONPs, which may interact with cell membrane proteins. The concentration of free Cu^{2+} ions in the aqueous solution around the CuONPs is, however, incredibly low due to its exceedingly poor solubility. Although the solubility of CuO varies with pH, it is around 3×10^5 M in pure water. This does not adequately explain the antibacterial action of CuONPs, which increases with concentration whereas CuO solubility remains constant at set pH and temperature [51].

4.3.2 Antifungal activity of CuONPs

Microorganisms can grow with harmful metals through certain mechanisms, which can result in the synthesis of nanoparticles as a consequence of reduction mechanisms. This field of nanobiotechnology is still in its infancy. 25 mL of $CuSO_4$. $5H_2O$ (10 mM) was combined with a cell-free supernatant after being heated at 100°C for 15 min. A color change was then noticed since these bacteria have a tendency to create enzymes that diminish the poisonous metal, resulting in the formation of CuO nanoparticles. Additional characteristics using XRD and TEM revealed that the average size of nanoparticle, which may have antibacterial properties, is 61.7 nm [52].

Copper nanoparticles (CuONPs) in the 10–190 nm range were produced using a fungal-mediated, cell-free extract of Trichoderma asperellum, and IR spectroscopy revealed that the amide and aromatic groups of the secondary metabolite act as an encapsulating or reducing agent. This nanoparticle was also used to study *in vitro* photothermal-induced therapy utilizing human lung carcinoma-A549 cancer cells. The fungus *Agaricus bisporus* produced CuNPs with a size range of 2–10 nm that exhibited antifungal activity [52].

4.3.3 Antioxidant activity

Complete Freund's adjuvant, a substance that mimics the progression of arthritic disease in humans, showed improved antioxidant enzymes and a decline in pro-inflammatory markers. In rats with induced arthritis, anti-inflammatory and anti-arthritic potentials were seen. Radical scavenging of 2,2-diphenyl-1-picryl-hydrazyl-hydrate (DPPH) is used to evaluate the DPPH radical-scavenging operation; the Shimada technique is applied. A 5 mL sample solution was mixed with an aliquot of DPPH methanol solution (20, 40, 60, and 80 g/mL) in 2 mL at various concentrations. After thoroughly shaking the mixture, it was let to stand at room temperature in the dark for 30 minutes. The absorbance was then measured in a spectrophotometer at 517 nm. The reaction mixture's lower absorbance showed a higher ability to scavenge free radicals [53].

4.4 Silver, gold, and copper nanoparticles used in biomedical applications

Silver, gold, and copper nanoparticles play crucial roles in advancing biomedical applications. Their unique properties, such as antimicrobial activity, optical properties, and biocompatibility, make them versatile tools in various fields within

Nanoparticle	Synthesis method	Size range	Biomedical applications	References
Silver-citrate nanoparticles	Chemical reduction	50 nm	Wound healing	[54]
Silver-chitosan nanoparticles	Green synthesis	80 nm	Drug delivery	[55]
Silver-alginate nanoparticles	Chemical reduction	30 nm	Imaging	[56]
Gold nanoparticles	Chemical reduction	5–26 nm	Photothermolysis of cancer cells and tumors, targeted delivery of drugs, and antigens	[57]
Curcumin-gold nanoparticles	Solvent casting technique	Nanoscale	Antibacterial activity	[58]
Cyclodextrin-capped gold nanoparticles	One-pot synthesis	Nanoscale	Drug delivery	[59]
Copper oxide	Green synthesis	Nanoscale	Antimicrobial agents, wound healing, drug delivery, cancer therapy, imaging agents	[60]
Copper sulfide	Solution casting method	Nanoscale	Photothermal therapy, imaging, drug delivery	[61]
Copper nanowires	Template-assisted electrodeposition method	Nanoscale	Sensors, antimicrobial coatings, wound healing, drug delivery	[62]
Copper nanoparticles	Green synthesis	Nanoscale	Antimicrobial coatings, wound healing, drug delivery, cancer therapy, imaging agents	[63]
Copper-gold	Chemical reduction	Nanoscale	Photothermal therapy, imaging, drug delivery	[64]
Copper-chitosan	Microwave-assisted method	Nanoscale	Wound healing, tissue engineering, drug delivery	[65]
Copper-graphene	Chemical method	Nanoscale	Antibacterial coatings, drug delivery, cancer therapy	[66]

Table 2.
Biomedical applications of silver, gold, and copper nanoparticles.

biomedicine. As research continues to unveil their potential benefits and address associated challenges, these nanoparticles could revolutionize diagnostics, therapy, and healthcare delivery, ultimately leading to improved patient outcomes and enhanced medical technologies.

Different forms of nanoparticles, their size ranges, and application in biomedical field have been presented in **Table 2**.

5. Challenges and prospective applications of green nanoparticles

The synthesis of green NPs and their applications is an area of significant interest due to the growing demand for sustainable and environmentally friendly technologies. Green NPs refer to NPs that are synthesized using eco-friendly and biocompatible methods, often utilizing natural sources and reducing the use of hazardous chemicals. Here are some challenges and perspectives for the synthesis of green NPs and their applications:

5.1 Challenges

a. **Standardization of synthesis methods:** Green nanoparticle synthesis methods often involve natural extracts, biomolecules, and bio-inspired approaches [67]. Standardizing these methods to ensure consistent nanoparticle size, shape, and properties can be challenging due to the variability of natural sources.

b. **Control over nanoparticle properties:** Achieving precise control over the size, shape, and composition of green nanoparticles can be difficult, which may impact their performance in various applications [68].

c. **Scale-up and production:** Many green synthesis methods are conducted in small batches [69]. Scaling up these methods to produce nanoparticles on a larger industrial scale while maintaining their green and sustainable nature presents technical and economic challenges.

d. **Characterization techniques:** Accurately characterizing green nanoparticles and understanding their properties require specialized techniques that might not be readily available or well-established.

e. **Stability and shelf life:** Ensuring the stability and shelf life of green nanoparticles can be challenging, as they may be susceptible to aggregation, degradation, or changes in their properties over time [70].

f. **Regulatory considerations:** New synthesis methods and applications for nanoparticles may raise regulatory concerns related to safety, toxicity, and environmental impact. Understanding and addressing these concerns is crucial for their successful implementation [71].

5.2 Perspectives

a. **Biomedical applications:** Green nanoparticles have promising applications in drug delivery, imaging, and targeted therapy due to their biocompatibility and reduced toxicity compared to traditional nanoparticles [72].

b. **Agriculture and food industry:** Green nanoparticles can be used for crop enhancement, pest control, and food preservation. They could replace conventional chemical-based approaches, reducing environmental impacts [73].

c. **Environmental remediation:** Green nanoparticles can be utilized for water purification, air filtration, and remediation of polluted environments, offering sustainable solutions to address pollution and contamination [74].

d. **Energy conversion:** Green nanoparticles can play a role in enhancing the efficiency of solar cells, catalytic processes, and energy storage devices, contributing to renewable energy technologies [75].

e. **Electronics and optoelectronics:** Green nanoparticles could find applications in flexible electronics, displays, and sensors due to their unique optical and electronic properties [76].

f. **Nanomedicine:** Green nanoparticles could be used in disease diagnosis, therapeutics, and imaging, opening up new possibilities for personalized medicine [77].

g. **Collaborative research:** Collaborations between scientists, engineers, biologists, chemists, and environmental experts are essential to address the multidisciplinary nature of green nanoparticle synthesis and applications [78].

h. **Education and awareness:** As green nanoparticles are relatively new, educating researchers, industries, and the public about their benefits, limitations, and responsible use is crucial for their successful integration into various sectors [79].

Overall, there are challenges associated with synthesizing green nanoparticles and applying them in various fields; the potential benefits in terms of sustainability, reduced environmental impact, and innovative applications make this area of research promising. Addressing the challenges through continued research, technological advancements, and collaborative efforts can lead to the development of practical and impactful green nanoparticle-based solutions.

6. Conclusion

Authors concluded that the research on the medicinal applications of Ag, Au, and CuNPs showed promising potential in various fields of medicine. Synthesis of Cu, Au, and AgNPs by using phytochemicals, microorganisms, and less toxic chemicals has been considered an effective and green approach. AgNPs possess strong antimicrobial properties and have been extensively studied for their potential in combating bacteria, viruses, and fungi. AgNPs have shown potential in cancer therapy due to their ability to induce apoptosis (cell death) in cancer cells and inhibit tumor growth. However, further research and clinical trials were needed to establish their efficacy and safety. AuNPs have been studied for their diagnostic and therapeutic applications in medicine. In cancer therapy, AuNPs have been explored as photothermal agents, using their ability to convert light into heat to destroy cancer cells selectively. CuNPs have also been studied for their antioxidant properties, indicating potential applications in reducing oxidative stress and inflammation-related conditions.

Acknowledgements

Authors are thankful to the manager of INTECHOPEN for providing opportunity to write a chapter for publication under book title "Advances in Green Chemistry: Catalysis and Renewable Feedstocks for Sustainable Chemical Synthesis."

Conflict of interest

Authors declared that they have no potential conflict of interest.

Author details

Faisal Nawaz*, Irum Jamil and Bushra Shakoor
Department of Chemistry, University of Wah, Wah, Pakistan

*Address all correspondence to: faisal.nawaz@uow.edu.pk

References

[1] Pryshchepa O, Pomastowski P, Buszewski B. Silver nanoparticles: Synthesis, investigation techniques, and properties. Advances in Colloid and Interface Science. 2020;**284**:102246

[2] Yang S-E, Liu P, Zhang Y-J, Guo Q-N, Chen Y-S. Effects of silver nanoparticles size and shape on light scattering. Optik. 2016;**127**(14):5722-5728

[3] Manivannan K, Cheng C-C, Anbazhagan R, Tsai H-C, Chen J-K. Fabrication of silver seeds and nanoparticle on core-shell Ag@ SiO2 nanohybrids for combined photothermal therapy and bioimaging. Journal of Colloid and Interface Science. 2019;**537**:604-614

[4] Bansal SA, Kumar V, Karimi J, Singh AP, Kumar S. Role of gold nanoparticles in advanced biomedical applications. Nanoscale Advances. 2020;**2**(9):3764-3787

[5] Klębowski B, Depciuch J, Parlińska-Wojtan M, Baran J. Applications of noble metal-based nanoparticles in medicine. International Journal of Molecular Sciences. 2018;**19**(12):4031

[6] Utreja P, Verma S, Rahman M, Kumar L. Use of nanoparticles in medicine. Current Biochemical Engineering. 2020;**6**(1):7-24

[7] Al-Khattaf FS. Gold and silver nanoparticles: Green synthesis, microbes, mechanism, factors, plant disease management and environmental risks. Saudi Journal of Biological Sciences. 2021;**28**(6):3624-3631

[8] Hekmati M, Hasanirad S, Khaledi A, Esmaeili D. Green synthesis of silver nanoparticles using extracts of Allium rotundum l, Falcaria vulgaris Bernh, and Ferulago angulate Boiss, and their antimicrobial effects in vitro. Gene Reports. 2020;**19**:100589

[9] Khalaj M, Kamali M, Costa MEV, Capela I. Green synthesis of nanomaterials-a scientometric assessment. Journal of Cleaner Production. 2020;**267**:122036

[10] Pal K, Chakroborty S, Nath N. Limitations of nanomaterials insights in green chemistry sustainable route: Review on novel applications. Green Processing and Synthesis. 2022;**11**(1):951-964

[11] Bruna T, Maldonado-Bravo F, Jara P, Caro N. Silver nanoparticles and their antibacterial applications. International Journal of Molecular Sciences. 2021;**22**(13):7202

[12] Savary S, Willocquet L, Pethybridge SJ, Esker P, McRoberts N, Nelson A. The global burden of pathogens and pests on major food crops. Nature Ecology & Evolution. 2019;**3**(3):430-439

[13] Shuping D, Eloff JN. The use of plants to protect plants and food against fungal pathogens: A review. African Journal of Traditional, Complementary and Alternative Medicines. 2017;**14**(4):120-127

[14] Baldrian P, Větrovský T, Lepinay C, Kohout P. High-throughput sequencing view on the magnitude of global fungal diversity. Fungal Diversity. 2022;**114**(1):539-547

[15] Salomoni R, Léo P, Montemor A, Rinaldi B, Rodrigues M. Antibacterial effect of silver nanoparticles

in Pseudomonas aeruginosa. Nanotechnology, Science and Applications. 2017;**10**:115-121

[16] Ismail M. Green synthesis and characterizations of copper nanoparticles. Materials Chemistry and Physics. 2020;**240**:122283

[17] Malhotra N, Ger T-R, Uapipatanakul B, Huang J-C, Chen KH-C, Hsiao C-D. Review of copper and copper nanoparticle toxicity in fish. Nanomaterials. 2020;**10**(6):1126

[18] Božić Cvijan B, Korać Jačić J, Bajčetić M. The impact of copper ions on the activity of antibiotic drugs. Molecules. 2023;**28**(13):5133

[19] Wang P, Yuan Y, Xu K, Zhong H, Yang Y, Jin S, et al. Biological applications of copper-containing materials. Bioactive Materials. 2021;**6**(4):916-927

[20] Manne R, Kumaradoss MMRM, Iska RSR, Devarajan A, Mekala N. Water quality and risk assessment of copper content in drinking water stored in copper container. Applied Water Science. 2022;**12**(3):27

[21] Chen J, Jiang Y, Shi H, Peng Y, Fan X, Li C. The molecular mechanisms of copper metabolism and its roles in human diseases. Pflügers Archiv-European Journal of Physiology. 2020;**472**:1415-1429

[22] Collins JF. Copper nutrition and biochemistry and human (patho) physiology. In: Advances in Food and Nutrition Research. Vol. 96. Academic Press; 2021. pp. 311-364

[23] Nieto-Maldonado A, Bustos-Guadarrama S, Espinoza-Gomez H, Flores-López LZ, Ramirez-Acosta K, Alonso-Nuñez G, et al. Green synthesis of copper nanoparticles using different plant extracts and their antibacterial activity. Journal of Environmental Chemical Engineering. 2022;**10**(2):107130

[24] Ermini ML, Voliani V. Antimicrobial nano-agents: The copper age. ACS Nano. 2021;**15**(4):6008-6029

[25] Crisan CM, Mocan T, Manolea M, Lasca LI, Tăbăran F-A, Mocan L. Review on silver nanoparticles as a novel class of antibacterial solutions. Applied Sciences. 2021;**11**(3):1120

[26] Bai X, Wang Y, Song Z, Feng Y, Chen Y, Zhang D, et al. The basic properties of gold nanoparticles and their applications in tumor diagnosis and treatment. International Journal of Molecular Sciences. 2020;**21**(7):2480

[27] Chen H, Zheng X, Chen Y, Li M, Liu K, Li X. Influence of copper nanoparticles on the physical-chemical properties of activated sludge. PLoS One. 2014;**9**(3):e92871

[28] Pareek V, Gupta R, Panwar J. Do physico-chemical properties of silver nanoparticles decide their interaction with biological media and bactericidal action? A review. Materials Science and Engineering: C. 2018;**90**:739-749

[29] Kart HH, Yildirim H, Kart SO, Çağin T. Physical properties of Cu nanoparticles: A molecular dynamics study. Materials Chemistry and Physics. 2014;**147**(1-2):204-212

[30] Garibo D, Borbón-Nuñez HA, de León JND, García Mendoza E, Estrada I, Toledano-Magaña Y, et al. Green synthesis of silver nanoparticles using Lysiloma acapulcensis exhibit high-antimicrobial activity. Scientific Reports. 2020;**10**(1):12805

[31] Elbeshehy EK, Elazzazy AM, Aggelis G. Silver nanoparticles synthesis

mediated by new isolates of Bacillus spp., nanoparticle characterization and their activity against bean yellow mosaic virus and human pathogens. Frontiers in Microbiology. 2015;**6**:453

[32] Javaid A, Oloketuyi SF, Khan MM, Khan F. Diversity of bacterial synthesis of silver nanoparticles. BioNanoScience. 2018;**8**:43-59

[33] Huang W, Chen S, Liu Y, Fu H, Wu G. The controlled synthesis of stable gold nanoparticles in quaternary ammonium ionic liquids by simple heating. Nanotechnology. 2010;**22**(2):025602

[34] Li Y, Wu T-Y, Chen S-M, Ali MA, AlHemaid FM. Green synthesis and electrochemical characterizations of gold nanoparticles using leaf extract of Magnolia kobus. International Journal of Electrochemical Science. 2012;**7**(12):12742-12751

[35] Singh PK, Kundu S. Biosynthesis of gold nanoparticles using bacteria. Proceedings of the National Academy of Sciences, India Section B: Biological Sciences. 2014;**84**:331-336

[36] Khan A, Rashid A, Younas R, Chong R. A chemical reduction approach to the synthesis of copper nanoparticles. International Nano Letters. 2016;**6**:21-26

[37] Alao II, Oyekunle IP, Iwuozor KO, Emenike EC. Green synthesis of copper nanoparticles and investigation of its anti-microbial properties. Advanced Journal of Chemistry-Section B. 2022;**4**(1):39-52

[38] Shantkriti S, Rani P. Biological synthesis of copper nanoparticles using Pseudomonas fluorescens. International Journal of Current Microbiology and Applied Sciences. 2014;**3**(9):374-383

[39] More PR, Pandit S, Filippis AD, Franci G, Mijakovic I, Galdiero M. Silver nanoparticles: Bactericidal and mechanistic approach against drug resistant pathogens. Microorganisms. 2023;**11**(2):369

[40] Mussin J, Giusiano G. Biogenic silver nanoparticles as antifungal agents. Frontiers in Chemistry. 2022;**10**:1023542

[41] Sathiya P, Geetha K. Fruit extract mediated synthesis of silver oxide nanoparticles using Dimocarpus longan fruit and their assesment of catalytic, antifungal, antioxidant and cytotoxic potentials. Inorganic and Nano-Metal Chemistry. 2021:1-11

[42] Jain N, Jain P, Rajput D, Patil UK. Green synthesized plant-based silver nanoparticles: Therapeutic prospective for anticancer and antiviral activity. Micro and Nano Systems Letters. 2021;**9**(1):5

[43] Alavi M, Hamblin MR, Martinez F, Kennedy JF, Khan H. Synergistic combinations of metal, metal oxide, or metalloid nanoparticles plus antibiotics against resistant and non-resistant bacteria. Micro Nano Bio Aspects. 2022;**1**(1):1-9

[44] Guy CS, Gibson MI, Fullam E. Targeting extracellular glycans: Tuning multimeric boronic acids for pathogen-selective killing of Mycobacterium tuberculosis. Chemical Science. 2019;**10**(23):5935-5942

[45] Alavi M, Kowalski R, Capasso R, Douglas Melo Coutinho H, Rose Alencar De Menezes I. Various novel strategies for functionalization of gold and silver nanoparticles to hinder drug-resistant bacteria and cancer cells. Micro Nano Bio Aspects. 2022;**1**(1):38-48

[46] Alavi M, Adulrahman NA, Haleem AA, Al-Râwanduzi ADH, Khusro A, Abdelgawad MA, et al.

Nanoformulations of curcumin and quercetin with silver nanoparticles for inactivation of bacteria. Cellular and Molecular Biology. 2021;**67**(5):151-156

[47] Mandhata CP, Sahoo CR, Padhy RN. Biomedical applications of biosynthesized gold nanoparticles from cyanobacteria: An overview. Biological Trace Element Research. 2022;**200**(12):5307-5327

[48] Fawad AA. Brief review on applications of gold and silver nanoparticles in drug delivery and DNA interaction. World Journal of Physics. 2023;**1**(1):5

[49] Mikhailova EO. Gold nanoparticles: Biosynthesis and potential of biomedical application. Journal of Functional Biomaterials. 2021;**12**(4):70

[50] Bhavyasree P, Xavier T. Green synthesised copper and copper oxide based nanomaterials using plant extracts and their application in antimicrobial activity. Current Research in Green and Sustainable Chemistry. 2022;**5**:100249

[51] Halbus AF, Horozov TS, Paunov VN. Strongly enhanced antibacterial action of copper oxide nanoparticles with boronic acid surface functionality. ACS Applied Materials & Interfaces. 2019;**11**(13):12232-12243

[52] Harishchandra BD, Pappuswamy M, Antony P, Shama G, Pragatheesh A, Arumugam VA, et al. Copper nanoparticles: A review on synthesis, characterization and applications. Asian Pacific Journal of Cancer Biology. 2020;**5**(4):201-210

[53] Thakar MA, Jha SS, Phasinam K, Manne R, Qureshi Y, Babu VH. X ray diffraction (XRD) analysis and evaluation of antioxidant activity of copper oxide nanoparticles synthesized from leaf extract of Cissus vitiginea. Materials Today: Proceedings. 2022;**51**:319-324

[54] Abdellatif AA, Rasheed Z, Alhowail AH, Alqasoumi A, Alsharidah M, Khan RA, et al. Silver citrate nanoparticles inhibit PMA-induced TNFα expression via deactivation of NF-κB activity in human cancer cell-lines, MCF-7. International Journal of Nanomedicine. 2020;**15**:8479-8493

[55] Oviya P, Francis AP, Mahalaxmi B. Fabrication of hemocompatible chitosan-biogenic silver nanocomposite for biomedical applications. Journal of Drug Delivery Science and Technology. 2023;**87**:104826

[56] Qamar S, Karim S, Aslam S, Jahangeer M, Nelofer R, Nadeem AA, et al. Alginate-based bio-nanohybrids with unique properties for biomedical applications. Starch-Stärke. 2022:2200100

[57] Herizchi R, Abbasi E, Milani M, Akbarzadeh A. Current methods for synthesis of gold nanoparticles. Artificial Cells, Nanomedicine, and Biotechnology. 2016;**44**(2):596-602

[58] Kolathupalayam Shanmugam B, Rajendran N, Arumugam K, Rangaraj S, Subramani K, Srinivasan S, et al. Curcumin loaded gold nanoparticles–chitosan/sodium alginate nanocomposite for nanotheranostic applications. Journal of Biomaterials Science, Polymer Edition. 2023;**34**(7):875-892

[59] Zhang NMY, Qi M, Wang Z, Wang Z, Chen M, Li K, et al. One-step synthesis of cyclodextrin-capped gold nanoparticles for ultra-sensitive and highly-integrated plasmonic biosensors. Sensors and Actuators B: Chemical. 2019;**286**:429-436

[60] Akintelu SA, Folorunso AS, Folorunso FA, Oyebamiji AK. Green

synthesis of copper oxide nanoparticles for biomedical application and environmental remediation. Heliyon. 2020;**6**(7):e04508

[61] Roy S, Rhim J-W. Gelatin-based film integrated with copper sulfide nanoparticles for active packaging applications. Applied Sciences. 2021;**11**(14):6307

[62] Mostafalu P, Sonkusale S. A high-density nanowire electrode on paper for biomedical applications. RSC Advances. 2015;**5**(12):8680-8687

[63] Din MI, Rehan R. Synthesis, characterization, and applications of copper nanoparticles. Analytical Letters. 2017;**50**(1):50-62

[64] Woźniak-Budych MJ, Langer K, Peplińska B, Przysiecka Ł, Jarek M, Jarzębski M, et al. Copper-gold nanoparticles: Fabrication, characteristic and application as drug carriers. Materials Chemistry and Physics. 2016;**179**:242-253

[65] Covarrubias C, Trepiana D, Corral C. Synthesis of hybrid copper-chitosan nanoparticles with antibacterial activity against cariogenic Streptococcus mutans. Dental Materials Journal. 2018;**37**(3):379-384

[66] Pavithra CL, Sarada BV, Rajulapati KV, Rao TN, Sundararajan G. A new electrochemical approach for the synthesis of copper-graphene nanocomposite foils with high hardness. Scientific Reports. 2014;**4**(1):4049

[67] Tidke PR, Gupta I, Gade AK, Rai M. Fungus-mediated synthesis of gold nanoparticles and standardization of parameters for its biosynthesis. IEEE Transactions on Nanobioscience. 2014;**13**(4):397-402

[68] Hui Y, Yi X, Hou F, Wibowo D, Zhang F, Zhao D, et al. Role of nanoparticle mechanical properties in cancer drug delivery. ACS Nano. 2019;**13**(7):7410-7424

[69] Tighe CJ, Cabrera RQ, Gruar RI, Darr JA. Scale up production of nanoparticles: Continuous supercritical water synthesis of Ce–Zn oxides. Industrial & Engineering Chemistry Research. 2013;**52**(16):5522-5528

[70] Zhang W. Nanoparticle aggregation: Principles and modeling. In: Nanomaterial: Impacts on Cell Biology and Medicine. 2014. pp. 19-43

[71] Fard JK, Jafari S, Eghbal MA. A review of molecular mechanisms involved in toxicity of nanoparticles. Advanced Pharmaceutical Bulletin. 2015;**5**(4):447

[72] McNamara K, Tofail SA. Nanoparticles in biomedical applications. Advances in Physics: X. 2017;**2**(1):54-88

[73] Khodakovskaya M, Marmiroli M. Polymeric nanoparticles for sustainable plant agriculture and food industry. Frontiers in Plant Science. 2023;**14**:1183938

[74] Rafeeq H, Hussain A, Ambreen A, Waqas M, Bilal M, Iqbal HM. Functionalized nanoparticles and their environmental remediation potential: A review. Journal of Nanostructure in Chemistry. 2022;**12**(6):1007-1031

[75] Kumar A, Choudhary P, Kumar A, Camargo PH, Krishnan V. Recent advances in plasmonic photocatalysis based on TiO2 and noble metal nanoparticles for energy conversion, environmental remediation, and organic synthesis. Small. 2022;**18**(1):2101638

[76] MacFarlane LR, Shaikh H, Garcia-Hernandez JD, Vespa M, Fukui T,

Manners I. Functional nanoparticles through π-conjugated polymer self-assembly. Nature Reviews Materials. 2021;**6**(1):7-26

[77] Perciani CT, Liu LY, Wood L, MacParland SA. Enhancing immunity with nanomedicine: Employing nanoparticles to harness the immune system. ACS Nano. 2020;**15**(1):7-20

[78] Grabbe S, Landfester K, Schuppan D, Barz M, Zentel R. Nanoparticles and the immune system: Challenges and opportunities. Nanomedicine. 2016;**11**(20):2621-2624

[79] Usmani A, Dash PP, Mishra A. Metallic nanoformulations: Green synthetic approach for advanced drug delivery. Materials Science: Advanced Composite Materials. 2018;**2**(2):1-4

Chapter 7

A Green Approach to Multifunctional Textiles: ZnO *In Situ* Synthesis Using Plant Waste Extracts and Gaseous Plasma

Anja Verbič, Gregor Primc, Martin Šala and Marija Gorjanc

Abstract

Introducing green technologies and minimising the use of synthetic and toxic compounds are the most important steps to overcome the environmental obstacles in textile finishing. Another essential requirement for a better life is the prevention of waste, which negatively impacts the environment, economy and human health. This chapter presents a comprehensive study on the development of a novel and environmentally friendly synthesis of zinc oxide (ZnO) using aqueous extracts from plant waste and gaseous plasma technology, directly (*in situ*) on cotton and polyethylene terephthalate (PET), to obtain fabrics with ultraviolet (UV)-protective and hydrophobic or hydrophilic properties. Plant waste from the food processing industry and invasive alien plants were used as a source of phytochemicals in *in situ* ZnO synthesis. ZnO is an inorganic compound that is widely used in various industries due to its multifunctional properties. It can exhibit UV-protective, antimicrobial, self-cleaning, hydrophobic and other properties when applied to textiles. The chapter investigates different methods and parameters to achieve the most optimised synthesis procedure to enable textiles with functional protective properties. It also discusses the importance of the selection of a plant-based reducing agent in ZnO synthesis, the use of gaseous plasma and its effect on polymer modification and assistance in ZnO synthesis.

Keywords: green synthesis, *in situ* synthesis, zinc oxide, ZnO, textile modification, plant waste extracts, gaseous plasma

1. Introduction

The textile industry is one of the biggest polluters of the environment. With growing consumerism and expansion of the textile industry, consumption of chemicals is also increasing. In the process of converting raw materials into finished textile products, as many as 3500 different chemicals are used, and approximately a tenth of those chemicals is confirmed to be hazardous to human health [1]. At the same time, the environmental awareness of the general population, industry and government organisations is growing. European Commission has accepted a commitment to become the first climate-neutral continent by 2050 [2]. By accepting the European Green Deal and Net-Zero Industry Act, the European Union (EU) has committed to ensure climate neutrality, clean technologies and to enable clean, healthy and affordable water, environment and food [2]. Negative consequences of mass production and use of chemicals have been recognised and the recommendations listed above are being encouraged for applications in different industries including textile. In the past years, the textile industry has been developing environmentally friendly processes, especially in areas where chemicals that pose a threat to human health and the environment are used. In addition to the ecological aspect, an essential challenge for the textile industry is to ensure competitiveness. That is why companies are moving towards developing multifunctional textile products with high-added value [3]. The modification of textile materials to improve functional properties (quality, comfort and protective properties) while preserving basic textile properties is the aim of many research articles. One of the ways in which new properties of a material can be imparted is by the application of nanoparticles. Most often, pre-prepared nanoparticles are applied to textiles because they have a known size and shape. At the same time, they often agglomerate and have poor adsorption to textile fibres [4]. The two-step process of applying pre-synthesised nanoparticles to textiles is called *ex situ* synthesis, as the particles are pre-synthesised and then deposited on the textile in a separate step. The opposite of *ex situ* is *in situ* synthesis, i.e., synthesis directly on the textile substrate. *In situ* synthesis is a more environmentally friendly alternative because it is a one-step process, which reduces the use of chemicals, water, time and energy. In addition, previous research has shown that it also enables uniform distribution and good adsorption of particles on textile substrates [5, 6]. Zinc oxide (ZnO) is an inorganic compound that has, in the last decade, been extensively used and researched in different industries, including textile, due to its multifunctional properties. In addition to its photocatalytic and semiconducting properties, functionalisation of textiles with ZnO particles exhibits ultraviolet (UV)-protective, antimicrobial and self-cleaning properties of the textile material [7]. Some authors also reported increased hydrophobic, flame-retardant, electrically conductive, thermally insulative and moisture managing properties of the textile material (**Figure 1**) [7].

While ZnO is found in nature in the form of the rare mineral zincite, it is mostly synthesised in laboratories. The chemical synthesis of ZnO in laboratories involves using various chemicals that act as reducing, stabilising, dispersing or binding agents in the process of synthesis or application to the textile substrate. Since the use of chemicals can be harmful to humans and the environment [8], the development of a new, environmentally friendly "green" process for the synthesis of ZnO is required.

In a typical chemical synthesis, ZnO is synthesised in the reaction between zinc salt (precursor) and reducing agent (usually NaOH or KOH). Hydroxyl (–OH) groups of the reducing agent are crucial for successful synthesis. They assist in the formation of intermediate zinc hydroxide, which is why a large amount of hydroxyl groups is required. During the drying process, zinc hydroxide is converted to ZnO [9–12]. In "green" synthesis procedures, chemical reducing agents can be replaced by ecologically acceptable alternatives, such as plant extracts, since plants are a great source of polyphenols, containing large amounts of hydroxyl groups. Additionally, considering that cotton contains hydroxyl group in its structure, our study also examined if the zinc salt and OH groups of the cotton alone are able to produce ZnO.

When reviewing the literature regarding the green *in situ* synthesis of ZnO on textiles, it was observed that there are a lot of articles that describe its processes as

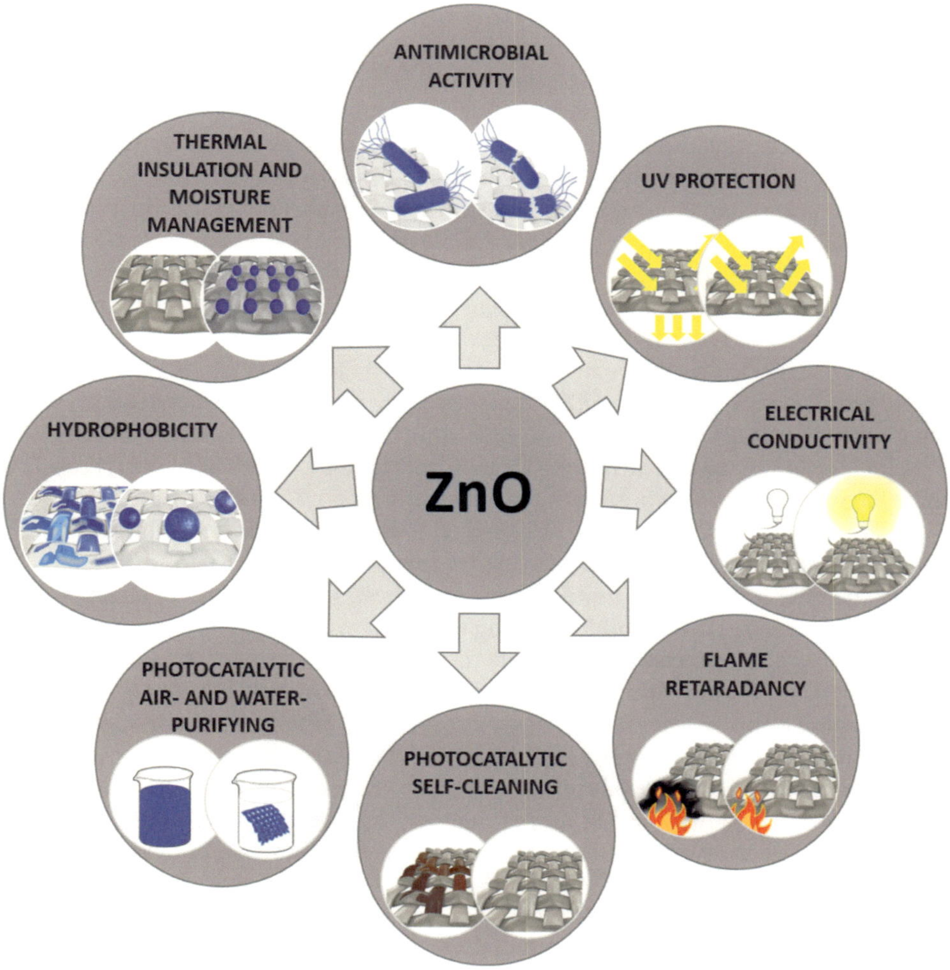

Figure 1.
Schematic presentation of the textile materials functionalities achieved with ZnO application. Reprinted with permission from [7].

green or *in situ*, but from the content it is clear that chemical reducing agents, which alone produce ZnO when combining them with zinc salt, are still included in the process or that harsh solvents were used to prepare plant extracts, or that the synthesis in fact does not take place directly on textiles. To our knowledge, there is only one research paper, where ZnO was synthesised directly on cotton using zinc salt and aqueous plant extract, without the addition of any other chemicals [13]. In the study by Aladpoosh et al., burnt *Seidlitzia rosmarinus* plant was used as a source of alkali. Authors reported good antimicrobial properties, tensile strength and crease recovery of the samples; however, the synthesis took place at an elevated temperature (90°C) for 90 minutes, meaning that the process was quite time- and energy-consuming. Furthermore, to achieve a truly environmentally friendly synthesis procedure, it has to be taken into account that the plant material used should not be grown specifically for the consumption as a plant extract for the synthesis. For the process to be considered as an example of a green circular economy, plant waste, which is usually discarded and burnt (such as food industry waste and invasive alien plant species), should be used to prepare the aqueous plant extracts, without the use of harsh solvents. It has also to be taken into account that the developed process should not be time- and energy-consuming.

Plasma treatment is an environmentally friendly technology in which an ionised gas is used to induce changes in the chemical and physical properties of textiles [3, 14]. Depending on the type of gas used and the plasma parameters applied, new reactive sites, cleaning, etching or polymerisation can be achieved. Compared to conventional processes, plasma treatment allows different effects to be achieved without using chemicals and water in shorter treatment times, affecting only the surface properties of the textile. Since the use of chemicals can have a negative impact on human health and the environment, the aim of this research was to develop an environmentally friendly process for the synthesis of ZnO directly (*in situ*) on cotton and polyethylene terephthalate (PET) fabrics, using oxygen plasma instead of chemical agents to increase the uptake of reagents, and using aqueous extracts of plant waste materials instead of conventional chemicals for ZnO synthesis, with the goal to achieve new functional properties of textiles.

Considering the existing gap of knowledge in the field of environmentally friendly *in situ* (direct) ZnO synthesis on textile substrates, this chapter presents a comprehensive study on the development of a novel, green procedure of ZnO synthesis directly on cotton and PET, using only zinc salt and plant waste aqueous extracts. Wood ash extract was used as alkali and aqueous extracts of food industry waste (avocado seed, avocado peel, green tea, pomegranate peel) or invasive alien plants (staghorn sumac leaves, staghorn sumac drupes, Japanese knotweed leaves) as reducing agents for the synthesis. The chapter addresses the importance of selecting an appropriate synthesis procedure, the order of synthesis solutions, different plant extracts as reducing agents and other parameters, which influence the successful formation of ZnO directly on textile substrates, with an emphasis on achieving the most optimised and environmentally friendly *in situ* ZnO synthesis procedure, resulting in textiles with functional protective properties. It also discusses the application of the optimised synthesis procedure from the cellulosic to synthetic textile material and the use of gaseous plasma and its effect on polymer modification and assistance in ZnO synthesis.

2. Experimental setup and methodology

2.1 Modification of textiles with *in situ* synthesised ZnO

2.1.1 Materials

Plain weaved, chemically bleached and mercerised cotton (Tekstina d.d., Ajdovščina, Slovenia) and de-lustered polyethylene terephthalate (PET) (Commerce, Ljubljana, Slovenia) were used as textile substrates. Zinc acetate dihydrate (Honeywell, Charlotte, NC, USA) and zinc nitrate hexahydrate (Sigma-Aldrich, Missouri, USA) were used as zinc precursors. Aqueous extracts of plant waste, i.e., food industry waste (avocado seed, avocado peel, green tea, pomegranate peel) or invasive alien plant species (staghorn sumac leaves, staghorn sumac drupes, Japanese knotweed leaves) were used as natural reducing agents. An aqueous extract of wood ash, a by-product of wood pellet heating system, was used as an alkaline medium. The extract preparation procedures are described in Verbič et al. 2021 [15].

2.1.2 Synthesis methods

Initially, four different synthesis methods were compared and examined to determine the optimal synthesis process [14]. All methods involved impregnation in three reaction solutions, i.e., alkali, zinc precursor and reducing agent. The first method (Method 1) involved only the successive impregnation of cotton in the reaction solutions. The second method (Method 2) involved padding the sample on two-cylinder foulard and drying in a continuous dryer after each impregnation. The third method (Method 3) involved ultrasonic shaking of the impregnation bath and drying of the samples in a continuous dryer after each impregnation. The fourth method (Method 4) involved drying the samples in a continuous oven after impregnation in each reaction solution (**Table 1**). Other synthesis parameters, such as concentration of reaction solutions, time and temperature of impregnation and drying, were standardised.

After determining the most optimal method, the order of the synthesis solutions was varied [16]. Two *in situ* synthesis procedures were compared. A procedure in which the impregnation in alkali was followed by a reducing agent and lastly zinc precursor, and a modified procedure in which the impregnation in alkali was followed by a zinc precursor and lastly a reducing agent. The experiments were carried out with different types and concentrations of zinc precursors and reducing agents [17]. The natural plant extracts used as reducing agents (avocado seed, avocado peel, green tea, pomegranate peel, staghorn sumac leaves, staghorn sumac drupes extract) were compared regarding their polyphenol content and antioxidant properties. The influence of using different reducing agents on the properties of *in situ* synthesised ZnO particles and fabric was examined.

2.2 Plasma modification of textiles

The procedure that proved to be optimal for the *in situ* synthesis of ZnO particles on cotton was then tested on PET fabric. Due to the hydrophobic properties of the PET fabric, the fabric was pre-treated with oxygen plasma for further investigations. We tested the effect of single (i), double (ii) and triple (iii) plasma treatments (before immersion in alkali (i), alkali and zinc precursor (ii), or alkali, zinc precursor and reducing agent (iii)) of the PET fabric [18] and lastly, compared the activation of the PET substrate by alkali and plasma [19].

Method no.	Cotton processing conditions according to the method
1	Alkali – Reducing agent – Zinc precursor – Drying 30 min at 100°C – Drying 5 min at 150°C
2	Alkali – Padding and drying 2 min at 100°C – Reducing agent – Padding and drying 5 min at 100°C – Zinc precursor – Padding and drying 5 min at 100°C – Drying 30 min at 100°C – Drying 5 min at 150°C
3	Alkali (sonicated) – Drying 2 min at 100°C – Reducing agent (sonicated) – Drying 5 min at 100°C – Zinc precursor (sonicated) – Drying 5 min at 100°C – Drying 30 min at 100°C – Drying 5 min at 150°C
4	Alkali – Drying 2 min at 100°C – Reducing agent – Drying 5 min at 100°C – Zinc precursor – Drying 5 min at 100°C – Drying 30 min at 100°C – Drying 5 min at 150°C

Table 1.
Description of different methods of in situ ZnO synthesis on cotton fabric [15].

2.3 Analysis of morphological, chemical and physical properties of textile samples

The morphological characteristics of the samples were examined by scanning electron microscopy (SEM) using JEOL JSM-6060 LV scanning electron microscope. Ultraviolet (UV) protection was determined by measuring the UV transmittance of the samples using a UV/visible (UV/Vis) spectrophotometer (Varian CARY 1E) following the AATCC test method 183-2000, from which the UV protection factor (UPF) was calculated. The samples' colour values (CIELAB) were measured on a reflectance spectrophotometer (Datacolor Spectraflash 600 PLUS-CT). Quantitative analysis of the ZnO distribution on the samples was carried out by inductively coupled plasma mass spectrometry (ICP-MS), using Agilent technologies 7900 ICP-MS (mass spectrometry) instrument, equipped with MicroMist glass concentric nebuliser and Peltier-cooled, Scott-type spray chamber. Elemental analysis of the surface of the samples was carried out by energy-dispersive spectroscopy (EDS) on a Quanta 650 Scanning Electron Microscope. Chemical changes on the textiles were determined by X-ray photoelectron spectroscopy (XPS) with Physical Electronics XPS instrument, X-ray powder diffraction (XRD) using Phillips PW3710 type diffractometer, X-ray fluorescence (XRF) using Thermo Fisher Scientific portable XRF analyser Niton XL5+ and Fourier transform infrared spectroscopy (FTIR) with Bruker Optics Vertex-70LS FTIR spectrometer. The mechanical properties of the samples were determined by measuring the tensile strength and elongation in accordance with standard ISO 13934–1:2013 on Instron 5567 strength tester. The chemical composition of the aqueous plant extracts was analysed by the determination of total phenolic compounds content (TPC) values in extracts by the Folin-Ciocalteu method (according to the AOAC 2017.13-2017 method). The antioxidant activity of the extracts and cotton samples was determined spectrophotometrically (PerkinElmer Lambda 850+ UV/Vis spectrophotometer) using DPPH (2,2-diphenyl-1-picrylhydrazyl) radical decolorisation. An optical assessment of the surface porosity was made using stereomicroscope (Leica Microsystems GmbH). The hydrophilicity or hydrophobicity of the samples was determined by measuring the water contact angles (WCAs) using a drop shape analyser DSA 100E (Krüss) and water absorption test was performed according to DIN 53924:2020 testing standard.

3. Results and discussion

This comprehensive study examines the environmentally friendly modification of textiles, with the aim of achieving new functional properties. We focused on the *in situ* (direct) synthesis of ZnO on textile substrates. Our main objective was to use aqueous extracts of plant waste instead of conventional chemicals, required for the successful synthesis of ZnO.

In our previous research in the field of classical chemical *in situ* synthesis of ZnO [4, 20], the influence of different reducing agents and their concentration on the formation of ZnO directly on cotton fabric was tested. It was found that the use of a lower concentration leads to uneven distribution of ZnO and the formation of agglomerates. Using a higher concentration of reducing agent resulted in larger ZnO layers with visible cracks. Furthermore, the use of a reducing agent with OH groups included in its chemical structure resulted in a higher amount of synthesised ZnO, and an even distribution of particles on cotton fibres, ensuing in higher UV-blocking

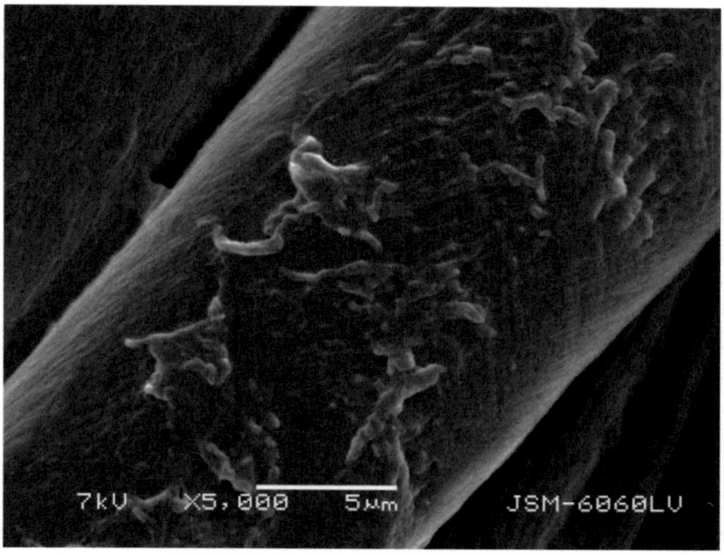

Figure 2.
SEM image of a cotton fabric functionalised with only zinc acetate. Reprinted with permission from [15].

ability. The results showed that an alkaline medium is essential for successfully synthesising ZnO on cotton and that a medium with OH groups incorporated into its chemical structure performs better as a reducing agent. As explained in the Introduction, the theory of ZnO synthesis states that OH groups are required to successfully synthesise ZnO from zinc salt. Since cotton contains a number of OH functional groups in its structure, we investigated whether the application of a zinc precursor and drying of the cotton fabric could be sufficient for the formation of ZnO [15]. But the result showed that this treatment forms a net-like structured layer on the fibres, presumed to be the crystalline form of zinc salt used [21]. There were no individual particles that could be ZnO on the fibres, meaning that the combination of cotton and zinc salt alone does not produce ZnO (**Figure 2**).

Furthermore, we investigated the formation of ZnO on cotton fabric using zinc acetate as a precursor and an aqueous extract of pomegranate peels as a reducing agent. Pomegranate peels contain high amounts of phenolic compounds, therefore, we were curious, if their numerous OH functional groups would be capable of forming ZnO. From the results [15], it is clear that an alkaline medium is necessary for successful synthesis of ZnO, as has been reported by others who have included NaOH or KOH in the synthesis reaction [20, 22, 23]. However, as the aim of our study was to exclude classical chemicals from the synthetic reaction, the alkaline medium was prepared from wood ash, a by-product of home heating system using wooden pellets as a fuel source. Phongarthit et al. [24] used wood ash in the *ex situ* synthesis of ZnO, and it proved to be an excellent alternative to NaOH when the calcination temperature of 900°C was used. An attempt of synthesising ZnO on cotton fabric was made by combining zinc precursor and alkali from wood ash and drying at 150°C, but the synthesis was unsuccessful. In our case, high calcination temperatures, such as 900°C, could not be used because the textile material would degrade. Thus, we attempted to achieve ZnO synthesis by using three reaction solutions, all prepared in water: an alkali prepared from wood ash (hereafter referred to as alkali), a reducing agent prepared from plant waste, i.e., plant food production waste or parts of invasive alien plant species, and a zinc precursor. The alkaline pre-treatment was used to achieve swelling of the cotton fibre, thereby increasing the number of free OH groups on the surface of the fibre and further boosting the binding possibility of various compounds. Initially, four synthesis methods were studied, which differed in the inclusion of drying the cotton fabric between impregnations in each reaction solution, padding the fabric on a two-cylinder foulard and drying, or ultrasonic shaking of the impregnation bath and drying (**Table 1**) [15].

All samples were prepared using the same concentrations of synthesis solutions (zinc precursor, pomegranate peel reducing agent and alkali). To maximise the energy efficiency of the process, the impregnation time for each reaction solution was limited to 1 minute at room temperature. Using Method 1 for the ZnO *in situ* synthesis, where drying the samples between the immersion into reaction solutions was not included, ZnO particles were successfully formed, and the fabric exhibited an extremely high UV protection factor (UPF 1000+). However, stiff, thick ZnO layers were formed, which peeled off the fabric's surface when handling. Since the sample was unfit for end-use, it was excluded from the further study. When the samples were padded on a two-roller foulard after each impregnation (Method 2), a significant amount of reaction compounds was removed from the sample, resulting in the lowest ZnO concentration and UPF value (**Figure 3**). Nevertheless, the sample still exhibited excellent UV protection (UPF 50+). The ultrasonication of the reaction solutions during the immersion, following by drying of the sample (Method 3), allowed a uniform distribution of ZnO particles on the fabric and also a higher ZnO content and UPF value than Method 1. However, the highest ZnO content and UPF values were achieved with Method 4, where the *in situ* ZnO synthesis included only drying of the fabric between the immersion in synthesis solutions. Drying the fabric between impregnations in the individual reaction solutions showed to be a necessary step to ensure an even distribution of ZnO throughout the substrate.

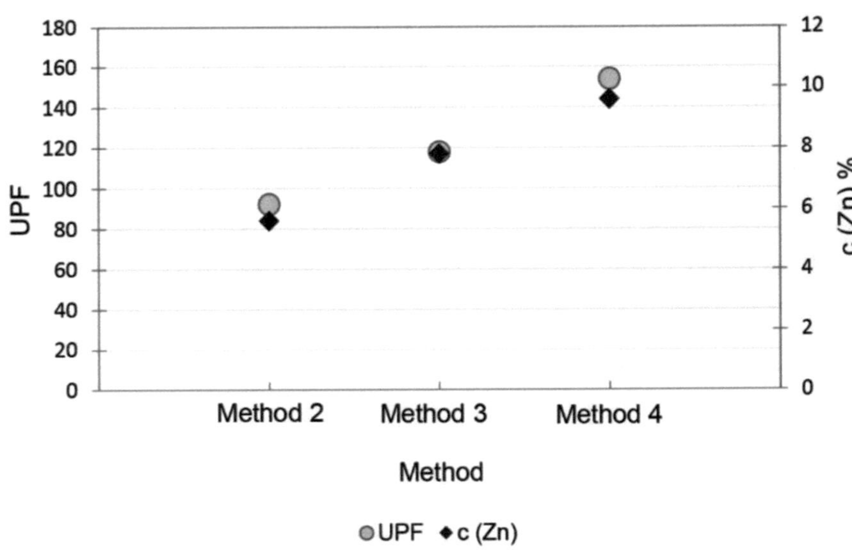

Figure 3.
Ultraviolet protection factor (UPF) values and zinc concentration (c (Zn)) of the samples where ZnO was in situ synthesised on cotton using methods 2–4.

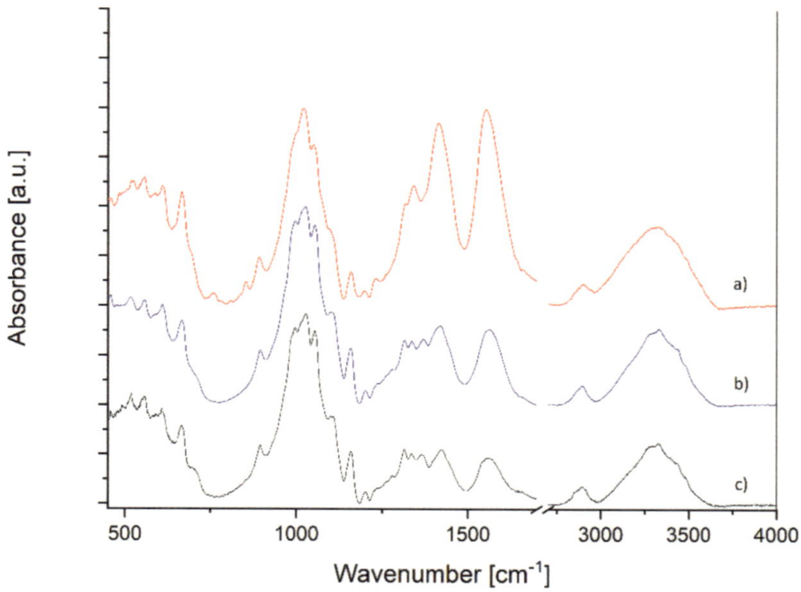

Figure 4.
IR spectra of cotton samples: (a) in situ synthesis of ZnO using method 4, (b) treated with zinc acetate precursor, (c) treated with pomegranate peel extract reducing agent. Reprinted with permission from [15].

The cotton sample where ZnO was *in situ* synthesised using Method 4 was analysed with FTIR analysis and compared with the cotton sample, treated only with zinc acetate and sample, treated with zinc acetate and pomegranate peel extract as reducing agent. The results showed that adding natural alkali (wood ash aqueous extract) in the first step of the synthesis is a necessary step which provides better adsorption of the synthesis reagents (**Figure 4**).

It was previously confirmed that plant extracts act as reducing agents and free radical scavengers due to their redox potential and high antioxidant ability caused by polyphenols [25–29]. Our results [15] showed a correlation between the antioxidant activity and values of total phenolic compounds content in the plant extracts used as reducing agents, while there was no correlation between antioxidant activity and the values of total flavonoid content. Nonetheless, the content of phenolic compounds varied between the extracts [30]. When comparing different plant extracts used as reducing agents (avocado seed and peel, green tea leaves and pomegranate peel), it was noticed that when using an extract with high content of phenolic compounds, a higher UPF value was achieved due to synthesised ZnO (very good or excellent UV blocking), and when using an extract with low values of phenolic compounds, lower UPF values were achieved (insufficient or minimal UV protection) (**Figure 5**). Furthermore, the antioxidant activity results showed that the samples functionalised with *in situ* synthesised ZnO using plant extracts as reducing agents have lower antioxidant activity than extract alone, because the phenolic compounds from the extract are consumed for binding on the fabric and in the reaction to form ZnO. Due to the consumption, a smaller amount of phenolic compounds is attached to the sample, influencing the measured value of the antioxidant activity. The samples with the highest ZnO content and highest UPF value had the lowest antioxidant activity due to the consumed phenolic compounds for the ZnO synthesis [15].

The cotton samples prepared with the extracts with the lowest TPC values had the lowest amount of synthesised ZnO particles (**Figure 5**). Those samples were also the least coated or had larger agglomerates of ZnO particles present, affecting the UV protection values (**Figures 5** and **6**) [30]. From the results, it was assumed that extracts with low concentrations of phenolic compounds could not react with the zinc precursor and further research was carried out to update the synthesis method.

Due to such different levels of polyphenols in the extracts (**Figure 5**), we had to modify the *in situ* synthesis method to allow the best possible reaction of polyphenols with zinc precursor on the textile substrate. The current synthesis method consisted of immersion in alkali, followed by a reducing agent and lastly, zinc precursor, meaning that polyphenols were first used to bind with the cellulose of the cotton fabric and then to form ZnO. That is why in the next step, the reaction solutions order was changed to alkali, zinc precursor, and lastly, reducing agent [16]. It was assumed that in this way, the reducing agents with low polyphenol content would have sufficient opportunity to react with the Zn precursor to form ZnO. The order of reaction solutions showed to have a significant effect on the formation of ZnO particles, reflecting in the protective properties of the functionalised fabric. With the modified method, the cotton fibres were entirely and uniformly coated with small ZnO particles that did not form agglomerates. The modified method enabled better reaction of the synthesis solutions, which resulted in higher zinc content (an increase from 3.2 to 8.4%) (**Figure 7**), as obtained by ICP-MS, offering

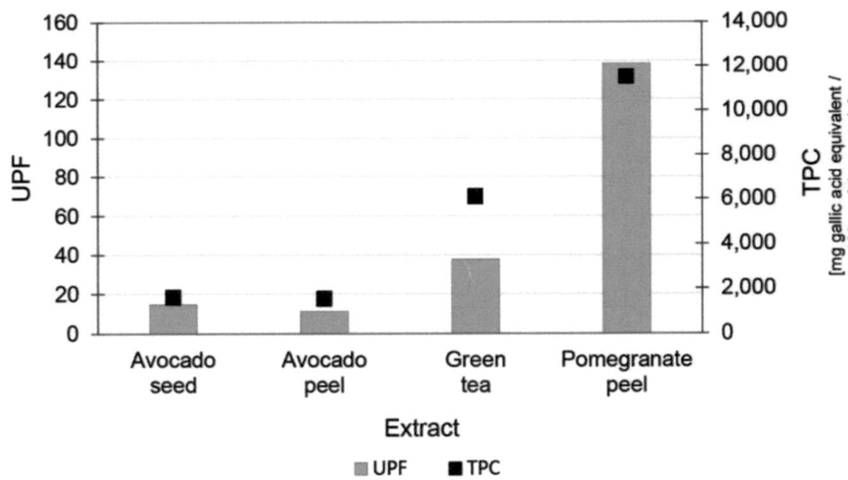

Figure 5.
Correlation between TPC value of the extract and UPF value of the cotton fabric with in situ synthesised ZnO using the same extract.

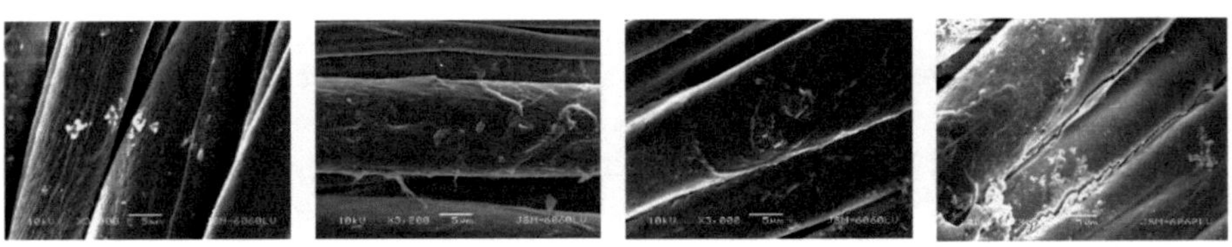

Figure 6.
Scanning electron microscopy (SEM) images of cotton samples functionalised using method 4, using (a) avocado seed, (b) avocado peel, (c) green tea and (d) pomegranate peel extract [30].

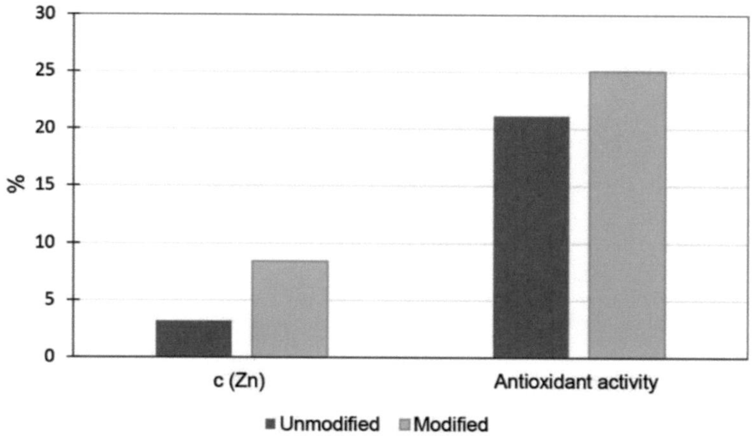

Figure 7.
Concentration of zinc (c(Zn)) and antioxidant activity (AA) of cotton samples where ZnO was in situ synthesised using unmodified and modified methods.

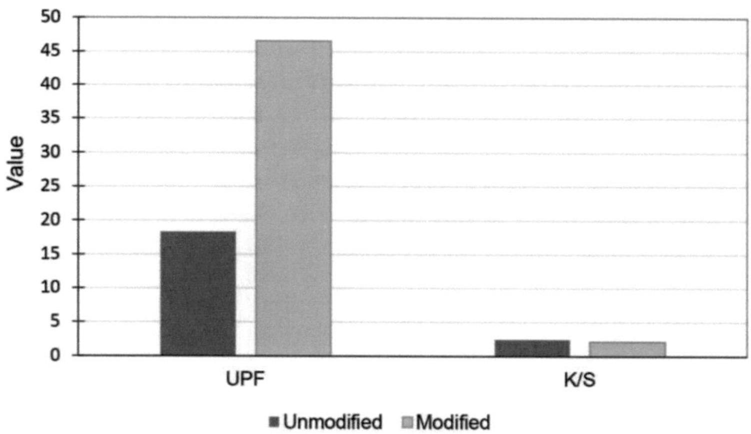

Figure 8.
Ultraviolet protection factor (UPF) and K/S values of cotton samples where ZnO was in situ synthesised using unmodified and modified methods.

improved fabric functionality. The morphology of the particles also changed since smaller, evenly distributed ZnO particles were produced. It is generally recognised that smaller particles with a higher specific surface area (surface-to-volume ratio) offer higher UV protection [22]. The modified procedure, where smaller and evenly distributed particles were synthesised, remarkably increased the UV-blocking ability of the samples (**Figure 8**). Since the colour of a sample can influence the ability to block UV radiation [31], we investigated for a possible correlation between the colour strength (K/S) and the achieved UPF values of the functionalised samples. The colour strength (K/S) results (**Figure 8**) showed that the samples have similar colour strength values (2.46 and 2.17) and at the same time very different UV-protective properties (UPF 18.33 and 46.57). A higher K/S value was measured for the sample that exhibited only minimal UV protection, while the K/S value was lower for the sample with very good UV protection. As a high K/S value means that the sample contains a high concentration of colourants [32], these results confirm

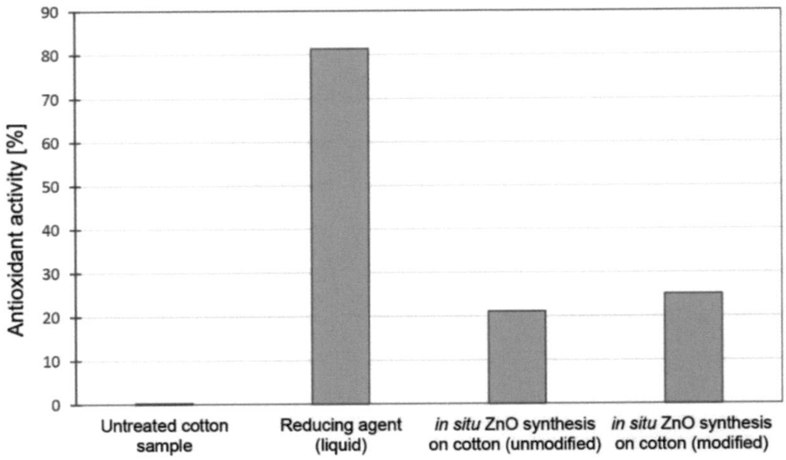

Figure 9.
Antioxidant activity of untreated cotton sample, liquid reducing agent and functionalised cotton samples using unmodified and modified synthesis methods [16].

that the UV protection achieved is a reflection of the synthesised ZnO particles and not due to the colour of the sample from the natural reducing agent.

Several published studies have confirmed [27–29, 33, 34] that the content of phenolic compounds in the natural extracts used as reducing agents for the synthesis of metal nanoparticles is crucial for their reducing ability, mainly due to their antioxidant properties, since antioxidants have good reducing capability. High content of phenolic compounds is therefore assumed to be crucial for the extract to perform as a reducing agent [30, 35]. Our results showed [15, 16, 30] that the extract used as a reducing agent has high values of antioxidant activity. When the extract was used for the *in situ* ZnO synthesis on cotton, the antioxidant value decreased in comparison with the extract alone (**Figure 9**).

Since phenolic compounds are consumed in the green synthesis process to form ZnO, based on the results of the UPF values, it would be expected that the sample prepared using the modified procedure (alkali → zinc precursor → reducing agent) would have a lower antioxidant activity value due to the lower content of polyphenolic compounds, which were consumed in the ZnO synthesis process. But in our study, the antioxidant activity values of the two samples were similar, or rather, using the unmodified procedure (alkali → reducing agent → zinc precursor), the antioxidant value was a little lower than when using the modified procedure (alkali → zinc precursor → reducing agent) [16]. This can be attributed to the fact that similar values of phenolic compounds were used for the synthesis, but in the case of the modified synthesis procedure, evenly distributed particles were formed (**Figure 10**), resulting in a higher UV protection (**Figure 8**). EDS analysis showed that the sample prepared with modified synthesis procedure contains a higher amount of zinc on the sample (**Table 2**) [16]. Furthermore, the colour measurements (K/S) of both samples were similar, indicating that a similar amount of phenolic compounds containing extract (reducing agent) was adsorbed on the samples (**Table 2**). Since better results were achieved with the modified ZnO *in situ* synthesis procedure, this procedure was used in further studies.

Next, we investigated how using different plant sources to prepare reducing agents can affect the successful ZnO synthesis [17]. As a source of phytochemicals for the

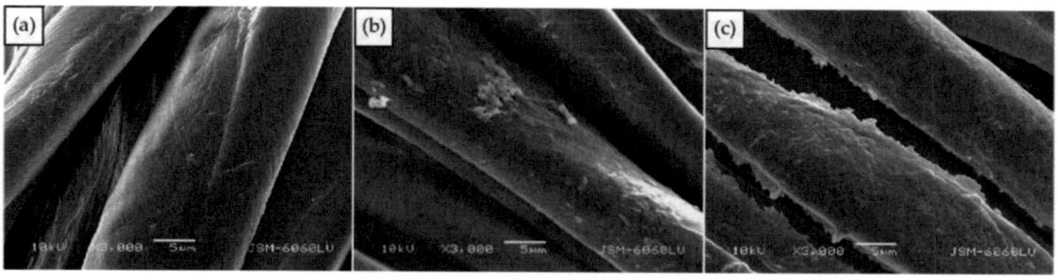

Figure 10.
SEM images of (a) untreated cotton sample, and cotton samples, functionalised with (b) unmodified ZnO in situ synthesis method and (c) modified ZnO in situ synthesis method. Reprinted with permission from [16].

Sample	Zn [%]	K/S (at 400 nm)
Untreated cotton	0	0.04
Unmodified ZnO *in situ* synthesis on cotton	3.2	2.45
Modified ZnO *in situ* synthesis on cotton	8.4	2.17

Table 2.
Zinc content (%) and K/S values of untreated cotton, and cotton samples functionalised with unmodified and modified ZnO in situ synthesis methods [16].

preparation of reducing agents, we used green tea leaves (GT), pomegranate peels (PG) and staghorn sumac leaves (SSL) and drupes (SSD). The selected waste plant sources were from the food waste group (green tea, pomegranate) and invasive alien plants in our local environment (staghorn sumac leaves and drupes). Furthermore, two different zinc precursors were used for the synthesis, zinc acetate and zinc nitrate. When the functionalised samples were prepared for analysis, it was observed that the samples synthesised with zinc nitrate as a precursor were very brittle. The mechanical properties of the samples were examined first by determining the breaking strength and elongation. We found that regardless of the type of reducing agent used, the samples where zinc nitrate was used in the synthesis process had significantly lower values of breaking strength and elongation (**Figure 11**). It is assumed that the mechanical properties of the samples were worsened due to the low pH value of the zinc precursor solution and because zinc nitrate has a low boiling point and it decomposes at around 125°C [36]. This temperature was exceeded in the last step of sample drying (5 minutes at 150°C in the laboratory oven). Also, the pH of the zinc nitrate solution was 3.64. Cellulose fibres are sensitive to acidic conditions, which can cause material degradation [37]. Since textile functionalisation processes should not significantly deteriorate the basic properties of the material, zinc nitrate was not suitable to be used as a precursor for *in situ* ZnO synthesis process on cotton, and was therefore excluded from further studies.

The samples where zinc acetate was used as a precursor for *in situ* ZnO synthesis were further analysed for UV transmittance, and UPF values were determined. Although all the functionalised fabrics achieved excellent UV protection (UPF 50+ according to the Australian/New Zealand standard AS/NZS 4399: 2017), the values varied between samples depending on the reducing agent used. The UPF values decreased in the following order: staghorn sumac leaves > pomegranate peel > green tea > staghorn sumac drupes (**Figure 12**) [17]. When preparing the samples for further analysis, it was observed that some samples showed hydrophilic, while others

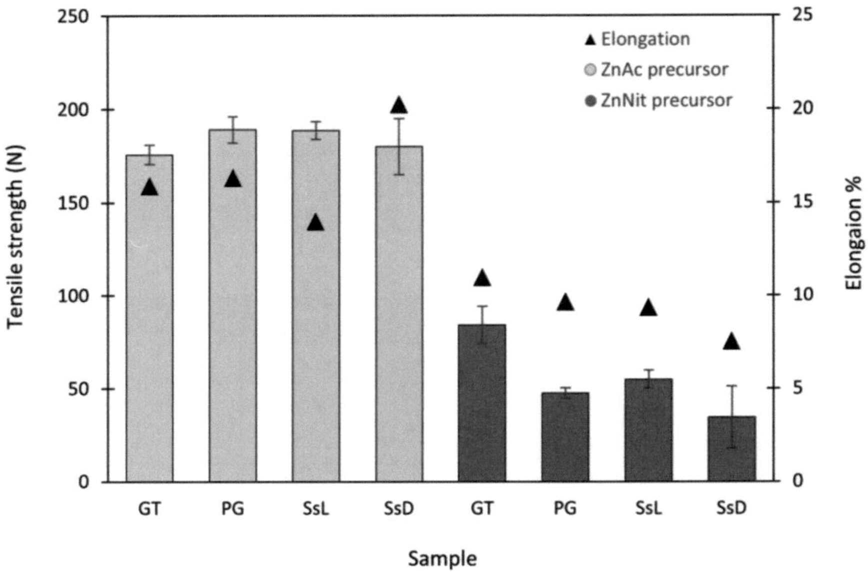

Figure 11.
Mechanical properties (tensile strength and elongation) of the cotton samples where in situ ZnO synthesis was performed using zinc acetate or zinc nitrate precursor using green tea (GT), pomegranate peel (PG), staghorn sumac leaves (SsL) and staghorn sumac drupes (SsD) reducing agent. Reprinted with permission from [17].

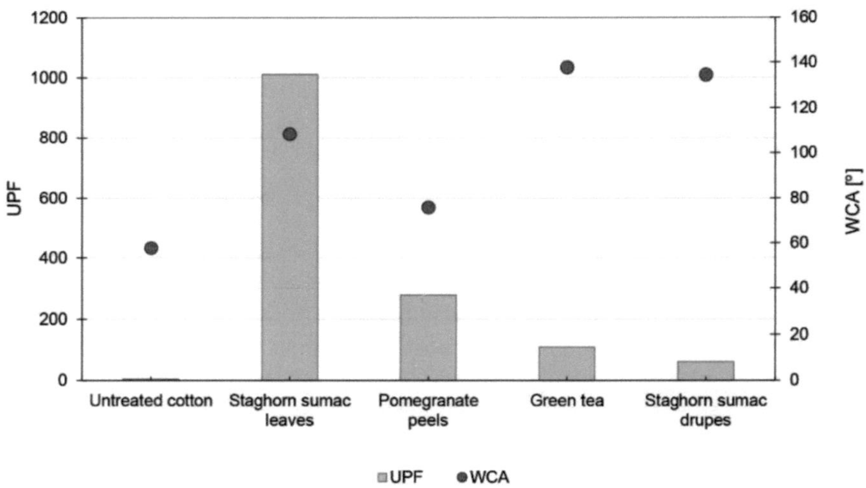

Figure 12.
Ultraviolet protection factor (UPF) values and water contact angles (WCAs) of the untreated cotton sample and cotton samples where ZnO was in situ synthesised using staghorn sumac leaves, pomegranate peel, green tea and staghorn sumac drupes [17].

showed hydrophobic properties, so the water contact angles of the samples were measured (**Figure 12**). It was observed that the samples with the highest UPF values (pomegranate peel, staghorn sumac leaves) had the lowest measured water contact angles and the samples with the lowest UPF values (green tea leaves, staghorn sumac drupes) had the highest water contact angles. In order to understand the connection between the extremely high UPF values and the different wettability of the fabrics, SEM, EDS, XPS, XRD, XRF and ICP-MS analyses were further performed [17].

DOI: http://dx.doi.org/10.5772/intechopen.1003131

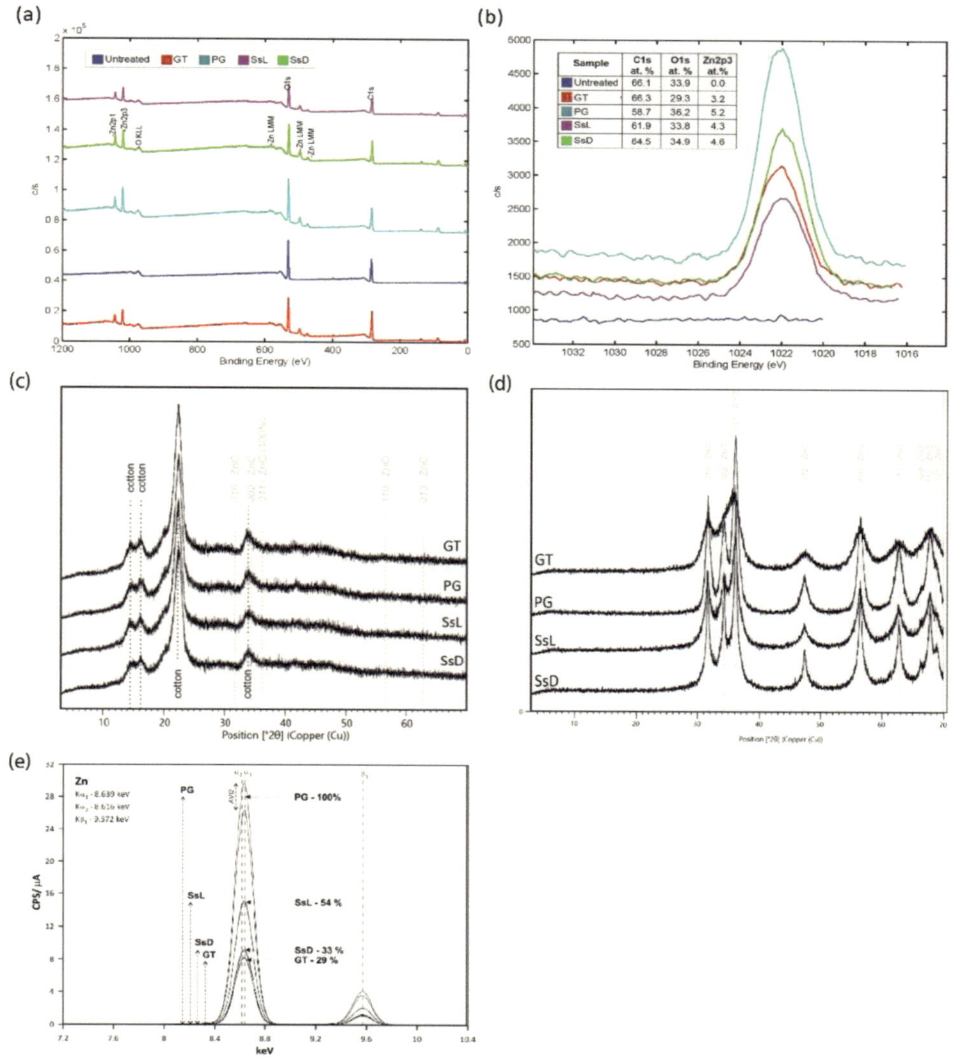

Figure 13.
XPS spectra of the untreated and in situ ZnO synthesis functionalised cotton samples (a, b), XRD spectra of in situ ZnO synthesis functionalised cotton samples (c), XRD spectra of ex situ synthesised ZnO (d) and XRF spectra of ex situ synthesised ZnO (e) using green tea (GT), pomegranate peel (PG), staghorn sumac leaves (SsL) and staghorn sumac drupes (SsD) reducing agent. Reprinted with permission from [17].

X-ray photoelectron spectroscopy (XPS) analysis confirmed that the identified compound on the spectrum is ZnO, with the presence of Zn $2p_1$, Zn $2p_3$ peaks, as well as the most intense Zn LMM Auger peak at 495 eV (**Figure 13a,b**) [17]. The XRD analysis results showed a weak intensity of the peaks characteristic of ZnO (**Figure 13c**), which could indicate that the synthesised ZnO is amorphous or that the ZnO content is too low to be detected. Although, various other researchers have confirmed that high temperature (above 150°C or even 300 or 600°C) is required for ZnO to the change from the amorphous to the crystalline form [38–41].

As cotton fabric is sensitive to such high temperatures, drying at temperatures this high cannot be included in the ZnO synthesis process directly on cotton. In order to prove that the synthesised particles are indeed ZnO, *ex situ* synthesised ZnO particles, without textile substrate, were prepared by the same procedure, which could be calcined at such high temperatures. Analyses (**Figure 13d**) of the prepared powders showed that crystalline ZnO was synthesised with distinct characteristic peaks corresponding to the mineral form of ZnO according to ICSD-98-000-9346. The intensity of the peaks varied depending on the type of extract used as a reducing agent. Using pomegranate peel or staghorn sumac leaves and drupes extract as reducing agents resulted in higher and narrower peaks, meaning that higher crystallinity of ZnO was achieved [42]. Moreover, no other peaks were observed in the spectra that were not specifically characteristic of ZnO, which means that the synthesised powder is pure ZnO with no impurities present. ICP-MS, XRF, EDS and UV/Vis results (**Figures 12** and **13d-e**) showed that the use of different extracts as reducing agents also influences the concentration of synthesised ZnO, which was highest when pomegranate peel extract and staghorn sumac leaves extract were used as reducing agents.

Various studies have indicated that surface morphology has a significant influence on the wettability of the substrate. It is generally accepted that increased surface roughness exhibits a more hydrophobic character [12]. Thi and Lee [43] describe that increased surface roughness contributes to more trapped air between the rough surface and water droplet, increasing the contact angle. In addition, when using nanoparticles to increase the surface roughness, the thickness of the nanoparticle layer is believed to have a major influence. A thicker ZnO layer is thought to increase the polar interactions with water, resulting in a lower contact angle and increased wettability. Our study [17] showed that the cotton samples on which ZnO was synthesised with pomegranate peel and staghorn sumac leaves extracts had the thickest layer of ZnO particles on the surface and, at the same time, had the lowest water contact angle values (**Figures 12** and **14**). These two extracts also had the highest TPC values (**Figure 15**). As the OH groups of the reducing agents were not entirely consumed

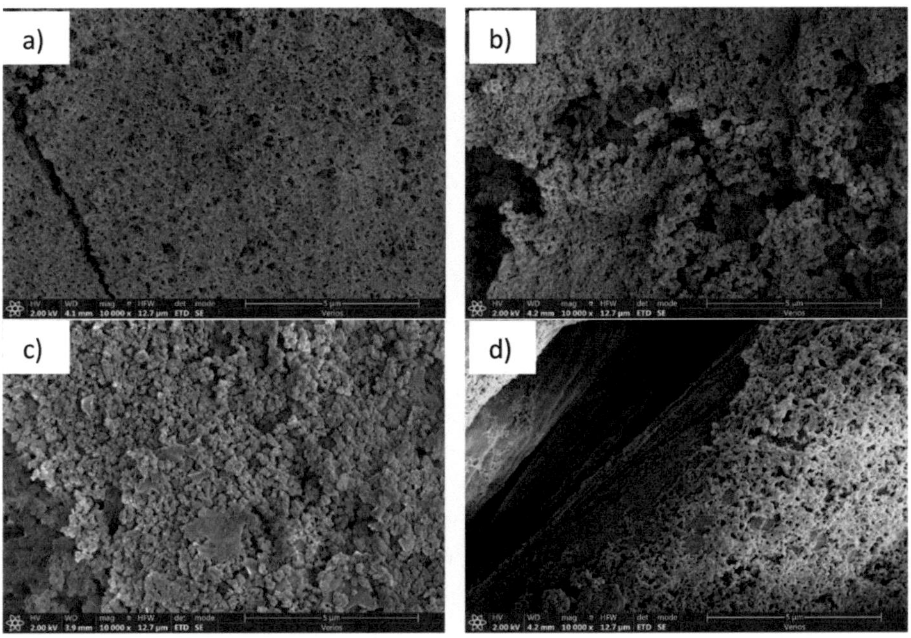

Figure 14.
SEM images of cotton fabric where in situ ZnO synthesis was performed using (a) green tea, (b) pomegranate peel, (c) staghorn sumac leaves and (d) staghorn sumac drupes extract. Reprinted with permission from [17].

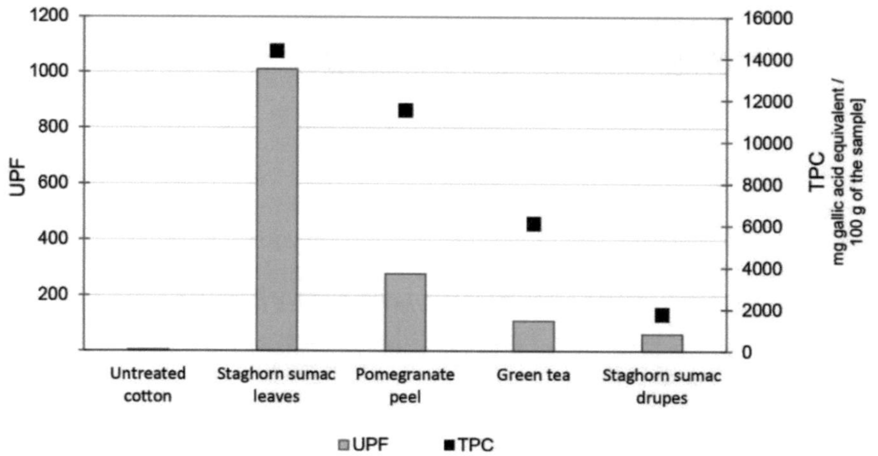

Figure 15.
Correlation between TPC values of the staghorn sumac leaves, pomegranate peel, green tea and staghorn sumac drupes extract and UPF values of untreated cotton fabric and cotton samples, where ZnO was in situ synthesised using the same extracts.

during the synthesis to form ZnO, they remained on the surface of the substrate, making these samples more hydrophilic. When green tea extract or staghorn sumac drupes extract, with lower TPC values, was used for ZnO synthesis, the OH groups were consumed during the synthesis process and the samples were more hydrophobic.

Green tea leaves, pomegranate peel, staghorn sumac leaves and drupes extracts were measured for TPC values (**Figure 15**). The highest values were measured in the staghorn sumac leaves extract (14,348 mg of gallic acid (GA) equivalent/100 g of the sample) and pomegranate peel extract (11,509 mg GA/100 g). The cotton samples on which ZnO was synthesised with these two extracts achieved the highest UV protection values and the lowest water contact angles. Meanwhile, the two samples prepared with extracts with lower TPC values (6103.2 mg GA/100 g for green tea leaves extract and 1769.4 mg GA/100 g for staghorn sumac drupes extract) achieved lower (but still excellent according to the AS/NZS 4399: 2017 standard) UV protection properties and had hydrophobic nature (high values of water contact angles). When green tea or staghorn sumac drupes extract was used, the polyphenolic compounds from the extract were mainly consumed for ZnO synthesis. The results of EDS, XRF and ICP-MS analyses [17] showed that the extracts containing a higher amount of total phenolic compounds were able to synthesise more ZnO particles. At the same time, many OH groups of the polyphenols were still retained on the surface of the ZnO, which affected the wettability of the samples. Therefore, textiles with different functional properties and their applicability can be designed by using different plant extracts as reducing agents in the ZnO synthesis process directly on textiles.

Since cotton and polyester are the most commonly used textile substrates, green *in situ* ZnO synthesis was also carried out on PET fabric. For the synthesis on PET, the same procedure as on cotton was used, i.e., impregnation in alkali, zinc precursor and reducing agent, with drying included between the impregnations. Pomegranate peel aqueous extract was used as the reducing agent, as excellent results were obtained with it in previous studies. Using the existing synthesis method, ZnO particles formed on the PET fabric and the UPF value of PET increased from 16.5 (untreated PET) to 177.9 (PET with *in situ* synthesised ZnO). The SEM analysis [18] showed that ZnO particles were not uniformly distributed over the sample and there were parts of the sample where ZnO was absent. Due to the hydrophobic nature of the PET fabric, the used reagents were not adsorbed uniformly on the sample, resulting in uneven synthesis of ZnO particles. For successful and uniform *in situ* synthesis of ZnO on PET fabric, the reactivity of the substrate needed to be increased. The classical chemical modification of PET to increase hydrophilicity is treatment in alkali [44]. Instead of using classical chemical treatment, we chose to increase the reactivity of our substrate by treating it with low-pressure inductively coupled plasma system, using oxygen as working gas. The low-pressure oxygen plasma treatment of PET causes partial hydrolysis of the ester bond and the formation of oxygen functional groups (COOH and OH) on the surface of the PET polymer fibres [45, 46]. The peaks characteristic of –C=O groups also increase [46–48]. The newly formed oxygen groups increase the hydrophilicity of PET, i.e., the WCA is decreased. In order to determine the optimal plasma treatment conditions, PET fabric was treated with oxygen plasma for 1, 4, 8 and 16 seconds, and the effect of plasma treatment duration on the achieved WCA was measured. The results are presented in **Figure 16**.

The untreated PET was poorly wettable with WCA of 72.1° ± 5.6°. The one-second plasma treatment lowered the sample's WCA to 22.1 ± 5.9°. While the individual values of the contact angles of the samples with the one-second plasma treatment are satisfactory, the standard deviation between the measured values remains high. The PET sample treated with plasma for 4 seconds became the most hydrophilic with an average contact angle of 11.2 ± 2.2°. This sample also achieved the lowest standard deviation between measurements, meaning it is the most uniformly wettable. As the plasma treatment time was extended, the contact angle values started to increase (to 26.9 ± 8.7° and 47.5 ± 3.2°). At the beginning of plasma treatment, the functionalisation of PET with oxygen-rich groups takes place, but with prolonged treatment time, etching begins to prevail. In addition, the samples that were plasma modified for 8 and 16 seconds were dimensionally deformed during processing and not perfectly flat, which is attributed to the fact that the PET

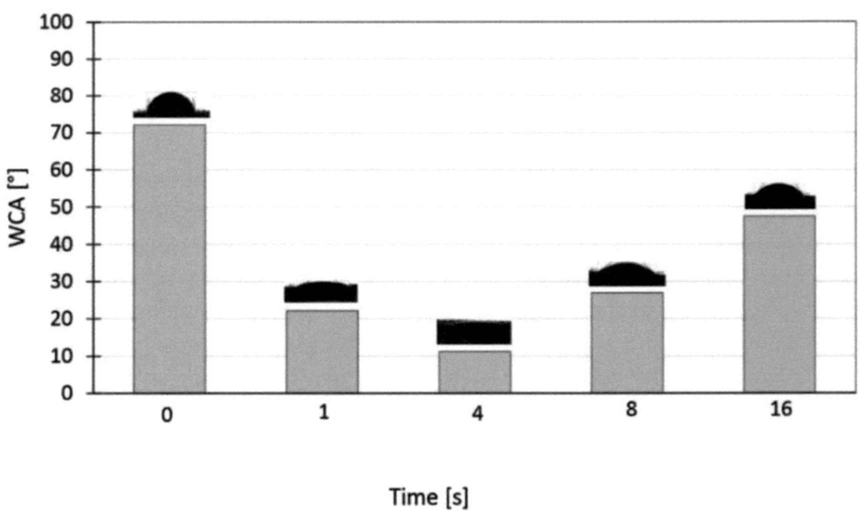

Figure 16.
Measurements and pictures of water contact angles (WCAs) of untreated PET and PET samples modified with different durations of oxygen plasma treatment [30].

fabric used was raw. Thus, 8- and 16-second plasma treatments were not appropriate for our sample at selected parameters.

Based on these results, we continued our study on *in situ* ZnO synthesis, where PET was pre-activated by a four-second plasma treatment. We decided to take an additional step of plasma treatment of the sample prior to impregnation in each reaction solution, i.e., prior to impregnation in alkali, prior to impregnation in alkali and zinc precursor, and prior to impregnation in alkali, zinc precursor and reducing agent. Therefore, single, double and triple plasma treatments were performed to increase reaction solution adsorption onto the PET substrate [18]. Due to the increased hydrophilicity of PET with plasma treatment, the reaction solutions were adsorbed uniformly, allowing ZnO particles to be synthesised evenly throughout the fabric. An even distribution of ZnO also increased UPF value of the fabric, which increased with the number of plasma treatments (UPF 218.6 > 295.2 > 303.7) (**Figure 17**).

Single, double and triple plasma-treated samples were analysed with EDS [18]. The analysis showed that the untreated PET sample contained high concentrations of carbon and oxygen, while no zinc was present on the sample. With single plasma activation before the *in situ* ZnO synthesis, the concentration of carbon decreased and the concentrations of oxygen and zinc increased. With increasing the number of plasma treatments (double and triple plasma activation), the values of zinc on the PET sample further increased, which confirms the activation of the PET substrate with oxygen plasma. The activation of PET substrate with oxygen plasma makes the sample more hydrophilic, further contributing to the formation of ZnO particles. As the number of plasma treatments increased, the ZnO particles were also more evenly distributed over the substrate, as confirmed by SEM analysis [18]. When PET is treated with oxygen plasma, a number of new reaction sites (oxygen functional groups) are created to which compounds from the reaction solutions can bind. This is reflected in the homogeneity of the coating and high UPF value, and also affects the colour values of the samples. As more compounds could be adsorbed on the sample, the K/S values of the samples increased with the number of plasma

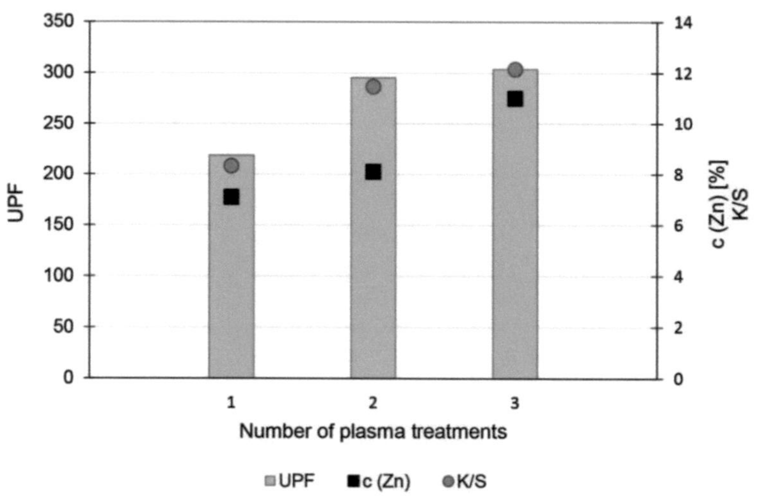

Figure 17.
Ultraviolet protection factor (UPF), K/S values and zinc concentration (c(Zn)) of the PET samples including single, double and triple plasma treatments.

treatments (from 8.32 to 12.14). The highest K/S value was measured for the sample that was plasma-treated also before impregnation in reducing agent. CIELAB colour measurements [18] showed that the samples treated with plasma become less red or greener (CIE a*), and also yellower (CIE b*) with the increasing number of plasma treatments.

As the primary goal of this comprehensive study was to make the *in situ* synthesis of ZnO on textiles as environmentally friendly as possible, and as alkaline PET pre-treatment has shown a favourable effect on ZnO formation, we further investigated whether plasma treatment has a sufficiently strong effect on polymer activation for successful ZnO synthesis without the use of alkali in the first step of the synthesis [19]. The results showed [19] that with both activation processes, a successful formation of ZnO particles can be achieved. Although, the water uptake analysis (**Figure 18**) showed that the sample activated in oxygen plasma had a better water uptake than the alkali-activated sample, suggesting that the plasma treatment produced more oxygen-rich functional groups on the PET substrate surface, which increased the hydrophilicity. Consequently, more reactive compounds could be adsorbed on the plasma-activated sample, resulting in a higher content and more even distribution of ZnO particles, ensuing in higher UPF values. Optical assessment of the fabric porosity of the sample showed that the sample where plasma was used to activate the substrate had a higher percentage of open area (higher fabric porosity) than the sample activated with alkali, but still achieved better protection against UV radiation, indicating that the high UPF value is due to the presence of synthesised ZnO particles.

The results show that both processes of PET activation are suitable, as both alkali and oxygen plasma enable the formation of hydrophilic active sites (COOH, and OH), due to the partial hydrolysis of the PET ester bond [45, 46]. Those are new adsorption sites where Zn ions of the zinc precursor can bind to. These sites present nucleation sites for the formation of ZnO particles. ZnO is synthesised in the third step of the synthesis process with the impregnation in a reducing agent,

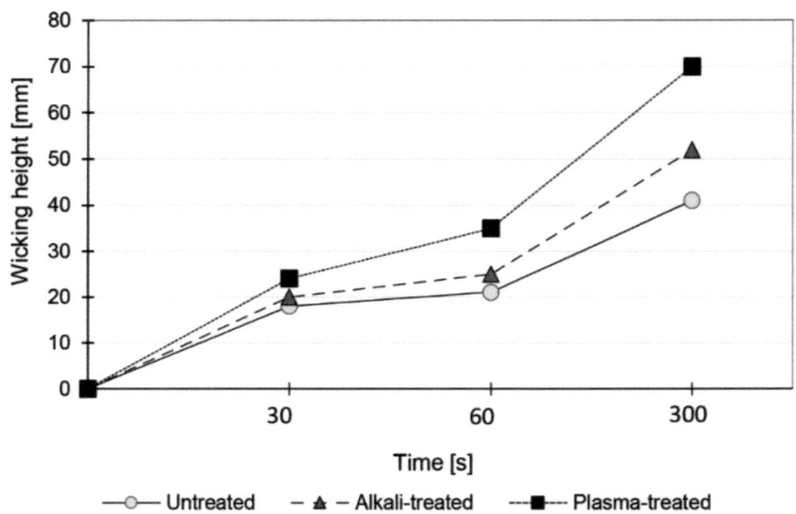

Figure 18.
Wicking height (mm) of the untreated, alkali-treated and plasma-treated PET samples, measured after 30, 60 and 300 seconds.

high in phenolic compounds. From the point of view of achieving a process that is as environmentally friendly as possible, it is also necessary to consider the preparation time and the consumption of reaction solutions (environmentally friendly alkali prepared in water) and to achieve as uniform distribution of ZnO particles on the fabric as possible, which makes plasma more appropriate for PET polymer activation.

4. Conclusions

This comprehensive study presents an important contribution to understanding the green synthesis of ZnO directly (*in situ*) on textile substrates. It addresses the importance of choosing the appropriate synthesis method, the selection and order of reagents for successful synthesis, and the influence of other parameters, such as the type of zinc precursor, the type and concentration of the "green" reducing agent, the influence of the pH value, the choice of the substrate and the plasma treatment on the chemical, morphological, mechanical and physical properties of the functionalised textiles. The findings provide a significant contribution to the development of functional protective textiles without the use of conventional chemicals which can be harmful to human health and the environment. Instead, it replaces conventional harmful chemicals with environmentally friendly aqueous extracts prepared from plant wastes, and by using oxygen plasma pre-treatment, thus contributing to solving the problem of plant waste and the development of protective textiles at the same time. Such modified textiles could be used for technical, protective or medical purposes in the future. The synthesis process is optimised in a way that low concentrations of reagents, low temperatures and short processing times are used, and could be transferred from the laboratory level to industrial applications.

Acknowledgements

This work was financially supported by the Slovenian Research Agency (Project J2-1720, Programme P2-0213, Programme P1-0034 and a grant for a doctoral student A. V.)

Author details

Anja Verbič[1,3], Gregor Primc[2], Martin Šala[3] and Marija Gorjanc[1*]

1 Faculty of Natural Sciences and Engineering, University of Ljubljana, Ljubljana, Slovenia

2 Jozef Stefan Institute, Ljubljana, Slovenia

3 National Institute of Chemistry, Ljubljana, Slovenia

*Address all correspondence to: marija.gorjanc@ntf.uni-lj.si

References

[1] Mohapatra P, Gaonkar O. An Overview of Toxic Chemicals in Textiles. New Delhi, India: Toxics Link; 2021. p. 41 Available from: https://www.researchgate.net/publication/358804949_AN_OVERVIEW_OF_TOXIC_CHEMICALS_IN_TEXTILES [Accessed: 2022-12-24]

[2] European Commission. A European green deal [Internet]. Available from: https://commission.europa.eu/strategy-and-policy/priorities-2019-2024/european-green-deal_en [Accessed: 2023-06-20]

[3] Gorjanc M. Plazemsko spremenjene površine bombaža za nanos nanosrebra [Doctoral dissertation]. University of Ljubljana: Faculty of Natural Sciences and Engineering. 2011

[4] Verbič A. Vpliv parametrov in-situ sinteze ZnO na bombažu za doseganje UV zaščite [Master's thesis]. University of Ljubljana: Faculty of Natural Sciences and Engineering. 2018

[5] Prasad V, Arputharaj A, Bharimalla AK, Patil PG, Vigneshwaran N. Durable multifunctional finishing of cotton fabrics by in situ synthesis of nano-ZnO. Applied Surface Science. 2016;**390**:936-940. DOI: 10.1016/j.apsusc.2016.08.155

[6] Bacciarelli-Ulacha A, Rybicki E, Matyjas-Zgondek E, Pawlaczyk A, Szynkowska MI. A new method of finishing of cotton fabric by in situ synthesis of silver nanoparticles. Industrial & Engineering Chemistry Research. 2014;**53**:4147-4155. DOI: 10.1021/ie4011113

[7] Verbič A, Gorjanc M, Simončič B. Zinc oxide for functional textile coatings: Recent advances. Coatings. 2019;**9**:550. DOI: 10.3390/coatings9090550

[8] Shah M, Fawcett D, Sharma S, Tripathy S, Poinern G. Green synthesis of metallic nanoparticles via biological entities. Materials. 2015;**8**:7278-7308. DOI: 10.3390/ma8115377

[9] Dhandapani P, Siddarth AS, Kamalasekaran S, Maruthamuthu S, Rajagopal G. Bio-approach: Ureolytic bacteria mediated synthesis of ZnO nanocrystals on cotton fabric and evaluation of their antibacterial properties. Carbohydrate Polymers. 2014;**103**:448-455. DOI: 10.1016/j.carbpol.2013.12.074

[10] Sundrarajan M, Ambika S, Bharathi K. Plant-extract mediated synthesis of ZnO nanoparticles using Pongamia pinnata and their activity against pathogenic bacteria. Advanced Powder Technology. 2015;**26**:1294-1299. DOI: 10.1016/j.apt.2015.07.001

[11] Yuvakkumar R, Suresh J, Nathanael AJ, Sundrarajan M, Hong SI. Novel green synthetic strategy to prepare ZnO nanocrystals using rambutan (Nephelium lappaceum L.) peel extract and its antibacterial applications. Materials Science and Engineering: C. 2014;**41**:17-27. DOI: 10.1016/j.msec.2014.04.025

[12] Kundu D, Hazra C, Chatterjee A, Chaudhari A, Mishra S. Extracellular biosynthesis of zinc oxide nanoparticles using Rhodococcus pyridinivorans NT2: Multifunctional textile finishing, biosafety evaluation and in vitro drug delivery in colon carcinoma. Journal of Photochemistry and Photobiology B: Biology. 2014;**140**:194-204. DOI: 10.1016/j.jphotobiol.2014.08.001

[13] Aladpoosh R, Montazer M. The role of cellulosic chains of cotton in biosynthesis of ZnO nanorods producing multifunctional properties: Mechanism, characterizations and features. Carbohydrate Polymers. 2015;**126**:122-129. DOI: 10.1016/j.carbpol.2015.03.036

[14] Gorjanc M, Mozetič M. Modification of Fibrous Polymers by Gaseous Plasma: Principles, Techniques and Applications. Saarbrücken: LAP Lambert Academic Publishing; 2014. p. 168

[15] Verbič A, Šala M, Jerman I, Gorjanc M. Novel green in situ synthesis of ZnO nanoparticles on cotton using pomegranate peel extract. Materials. 2021;**14**:4472. DOI: 10.3390/ma14164472

[16] Verbič A, Brenčič K, Primc G, Gorjanc M. Importance of protocol design for suitable green in situ synthesis of ZnO on cotton using aqueous extract of japanese knotweed leaves as reducing agent. Forests. 2022;**13**:143. DOI: 10.3390/f13020143

[17] Verbič A, Brenčič K, Dolenec M, Primc G, Recek N, Šala M, et al. Designing UV-protective and hydrophilic or hydrophobic cotton fabrics through in-situ ZnO synthesis using biodegradable waste extracts. Applied Surface Science. 2022;**599**:153931. DOI: 10.1016/j.apsusc.2022.153931

[18] Verbič A, Brenčič K, Primc G, Mozetič M, Gorjanc M. Eco-friendly In situ ZnO synthesis on PET fabric using oxygen plasma and plant waste. Coatings. 2022;**12**:537. DOI: 10.3390/coatings12040537

[19] Verbič A, Brenčič K, Primc G, Mozetič M, Šala M, Gorjanc M. Plasma-assisted green synthesis of ZnO directly on polyethylene terephthalate fabric. Fibers and Polymers. 2023;**24**:1-12. DOI: 10.1007/s12221-023- 00059-1

[20] Verbič A, Šala M, Gorjanc M. The influence of in situ synthesis parameters on the formation of ZnO nanoparticles and the UPF value of cotton fabric. Tekstilec. 2018;**61**:280-288. DOI: 10.14502/Tekstilec2018.61.280-288

[21] Hu XL, Masuda Y, Ohji T, Kato K. Rapid low-temperature synthesis of porous ZnO nanoparticle film by self-hydrolysis technique. Key Engineering Materials. 2010;**445**:123-126. DOI: 10.4028/www.scientific.net/kem.445.12

[22] Becheri A, Dürr M, Lo Nostro P, Baglioni P. Synthesis and characterization of zinc oxide nanoparticles: Application to textiles as UV-absorbers. Journal of Nanoparticle Research. 2008;**10**:679-689. DOI: 10.1007/s11051- 007-9318-3

[23] Yadav A, Prasad V, Kathe AA, Raj S, Yadav D, Sundaramoorthy C, et al. Functional finishing in cotton fabrics using zinc oxide nanoparticles. Bulletin of Materials Science. 2006;**29**:641-645. DOI: 10.1007/s12034-006-0017-y

[24] Phongarthit K, Amornpitoksuk P, Suwanboon S. Synthesis, characterization, and photocatalytic properties of ZnO nanoparticles prepared by a precipitation-calcination method using a natural alkaline solution. Materials Research Express. 2019;**6**:045501. DOI: 10.1088/2053-1591/aaf8db

[25] Kähkönen MP, Hopia AI, Vuorela HJ, Rauha JP, Pihlaja K, Kujala TS, et al. Antioxidant activity of plant extracts containing phenolic compounds. Journal of Agricultural and Food Chemistry. 1999;**47**:3954-3962. DOI: 10.1021/jf990146l

[26] Al-Rimawi F, Abu-Lafi S, Abbadi J, Alamarneh AA, Sawahreh RA, Odeh I. Analysis of phenolic and flavonoids of wild Ephedra alata plant extracts by LC/PDA and LC/MS and their

antioxidant activity. African Journal of Traditional, Complementary and Alternative Medicines. 2017;**14**:130-141. DOI: 10.21010/ajtcam.v14i2.14

[27] Senthilkumar SR, Sivakumar T. Green tea (Camellia sinensis) mediated synthesis of zinc oxide (ZnO) nanoparticles and studies on their antimicrobial activities. International Journal of Pharmacy and Pharmaceutical Sciences. 2014;**6**:461-465

[28] Selim YA, Azb MA, Ragab I, Abd El-Azim MHM. Green synthesis of zinc oxide nanoparticles using aqueous extract of Deverra tortuosa and their cytotoxic activities. Scientific Reports. 2020;**10**:1-9. DOI: 10.1038/s41598- 020-60541-1

[29] Hashemi S, Asrar Z, Pourseyedi S, Nadernejad N. Green synthesis of ZnO nanoparticles by olive (Olea europaea). IET Nanobiotechnology. 2016;**10**:400-404. DOI: 10.1049/iet-nbt.2015.0117

[30] Verbič A. In situ sinteza nanodelcev ZnO s pomočjo fitokemikalij in plinske plazme za doseganje večfunkcionalnih lastnosti tekstilij [Doctoral dissertation]. University of Ljubljana: Faculty of Natural Sciences and Engineering. 2023

[31] Gorjanc M, Mozetič M, Vesel A, Zaplotnik R. Natural dyeing and UV protection of plasma treated cotton. The European Physical Journal D. 2018;**72**:1-6. DOI: 10.1140/epjd/e2017-80680-9

[32] Gorjanc M, Savić A, Topalić-Trivunović L, Mozetič M, Zaplotnik R, Vesel A, et al. Dyeing of plasma treated cotton and bamboo rayon with Fallopia japonica extract. Cellulose. 2016;**23**:2221-2228. DOI: 10.1007/s10570-016-0951-9

[33] Nadagouda MN, Iyanna N, Lalley J, Han C, Dionysiou DD, Varma RS. Synthesis of silver and gold nanoparticles using antioxidants from blackberry, blueberry, pomegranate, and turmeric extracts. ACS Sustainable Chemistry & Engineering. 2014;**2**(7):1717-1723. DOI: 10.1021/sc500237k

[34] Gour A, Jain NK. Advances in green synthesis of nanoparticles. Artificial Cells, Nanomedicine, and Biotechnology. 2019;**47**:844-851. DOI: 10.1080/21691401.2019.1577878

[35] Majhi K, Let M, Kabiraj A, Sarkar S, Halder U, Dutta B, et al. Metal recovery using nanobiotechnology. In: Nanobiotechnology: Microbes and Plant Assisted Synthesis of Nanoparticles, Mechanisms and Applications. Amsterdam, NL: Elsevier; 2021. pp. 283-301

[36] American Elements. Zinc nitrate hexahydrate [Internet]. Available from: https://www.americanelements.com/zinc-nitrate-hexahydrate-10196-18-6 [Accessed: 2023-07-25]

[37] Huang YB, Fu Y. Hydrolysis of cellulose to glucose by solid acid catalysts. Green Chemistry. 2013;**15**:1095-1111. DOI: 10.1039/c3gc40136g

[38] Ramesh TN, Madhu TL. Thermal decomposition studies of layered metal hydroxynitrates (metal: Cu, Zn, Cu/Co, and Zn/Co). International Journal of Inorganic Chemistry. 2015;**2015**:11. Article ID: 536470. DOI: 10.1155/2015/536470

[39] Al-Hada NM, Saion EB, Shaari AH, Kamarudin MA, Flaifel MH, Ahmad SH, et al. A facile thermal-treatment route to synthesize ZnO nanosheets and effect of calcination temperature. PLoS One. 2014;**9**:e103134. DOI: 10.1371/journal.pone.0103134

[40] Parra MR, Haque FZ. Aqueous chemical route synthesis and the effect

of calcination temperature on the structural and optical properties of ZnO nanoparticles. Journal of Materials Research and Technology. 2014;**3**:363-369. DOI: 10.1016/j.jmrt.2014.07.001

[41] Karthik S, Siva P, Balu KS, Suriyaprabha R, Rajendran V, Maaza M. Acalypha indica–mediated green synthesis of ZnO nanostructures under differential thermal treatment: Effect on textile coating, hydrophobicity, UV resistance, and antibacterial activity. Advanced Powder Technology. 2017;**28**:3184-3194. DOI: 10.1016/J.APT.2017.09.033

[42] Soosen Samuel M, Bose L, George KC. Optical properties of ZnO nanoparticles. Academic Review. 2009;**16**:57-65

[43] Thi VHT, Lee BK. Development of multifunctional self-cleaning and UV blocking cotton fabric with modification of photoactive ZnO coating via microwave method. Journal of Photochemistry and Photobiology A: Chemistry. 2017;**338**:13-22. DOI: 10.1016/j.jphotochem.2017.01.020

[44] Dave J, Kumar R, Srivastava HC. Studies on modification of polyester fabrics I: Alkaline hydrolysis. Journal of Applied Polymer Science. 1987;**33**:455-477

[45] Ibrahim NA, Eid BM, Khalil HM, Almetwally AA. A new approach for durable multifunctional coating of PET fabric. Applied Surface Science. 2018;**448**:95-103. DOI: 10.1016/j.apsusc.2018.04.022

[46] Kamel MM, El Zawahry MM, Helmy H, Eid MA. Improvements in the dyeability of polyester fabrics by atmospheric pressure oxygen plasma treatment. The Journal of the Textile Institute. 2011;**102**:220-231. DOI: 10.1080/00405001003672366

[47] Gupta B, Hilborn J, Hollenstein CH, Plummer CJG, Houriet R, Xanthopoulos N. Surface modification of polyester films by RF plasma. Journal of Applied Polymer Science. 2000;**78**:1083-1091. DOI: 10.1002/1097-4628(20001031)78:53.0.CO;2-5

[48] Mihailović D, Šaponjić Z, Molina R, Puač N, Jovancić P, Nedeljković J, et al. Improved properties of oxygen and argon RF plasma-activated polyester fabrics loaded with TiO_2 nanoparticles. ACS Applied Materials & Interfaces. 2010;**2**:1700-1706. DOI: 10.1021/am100209n

Section 4

Strategy S: Energy and Resource Sustainability

Chapter 8

Green Chemistry of Minimal-Protection Solid-Phase Peptide Synthesis

Yi Yang

Abstract

The conventional SPPS (solid-phase peptide synthesis) requires protecting the nucleophilic side chains of amino acids to prevent undesired modifications. A large volume of TFA (trifluoroacetic acid) is generally needed to remove these protecting groups post the peptide assembly. Such a process significantly lowers the productivity of the subject peptide manufacturing and is deemed contradictory to green chemistry concepts. Enabling the side-chain unprotected amino acid couplings should break through such a bottleneck in peptide production and drastically increase productivity. This aspiration creates the concept of MP-SPPS (Minimal-Protection Solid-Phase Peptide Synthesis), accomplished in peptide API (Peptide Z) manufacturing. Side-chain unprotected arginine and tyrosine have been successfully incorporated into the target peptide chain on solid supports. The target peptide Z could be readily obtained by treating the peptidyl resin with a diluted acid solution, that is, 10% TFA/TFT (trifluorotoluene), and precipitating the solid product in a radically reduced volume of anti-solvent. The MP-SPPS process achieves a 5.3-fold increase in peptide Z manufacturing and complies with the philosophy of green chemistry.

Keywords: Minimal-Protection Solid-Phase Peptide Synthesis, green chemistry, productivity, trifluorotoluene, unprotected arginine, unprotected tyrosine

1. Introduction

To date, green chemistry has focused on improving the safety and environmental sustainability of the manufacturing process, including the organic solvents commonly used in peptide manufacturing. Due to the inherent advantages of their delicate balance to solubilize various compounds, including amino acid building blocks, coupling reagents, reactive intermediates, and by-products, and to properly swell peptidyl resins, DMF (dimethylformamide) and NMP (*N*-methylpyrrolidone) are routinely utilized as default organic solvents for SPPS (Solid-Phase Peptide Synthesis). Moreover, DMF and NMP lower the occurrence of prevalent side reactions such as amino acid racemization, aspartimide formation, DKP formation, *N*-terminus capping, and premature peptide cleavage [1]. Despite these merits, DMF and NMP are not deemed green solvents, and alternative organic solvents for peptide synthesis have been widely pursued [2]. No green solvents to date can ensure high purity and excellent yield for any peptide synthesis, which represents one main impediment to their regular utilization for peptide manufacturing [3]. For instance, the target peptide sequence and the inherent side reactions induced during peptide synthesis dictate the choice of the optimal solvent. Consequently, lowering the amount of solvent used in SPPS and developing green solvent applications in peptide synthesis deserves in-depth investigation and synergic contributions from academia and industrial domains. Many insightful papers have been published describing the green solvent performances in peptide chemistry [4–7]. The utilization of authentic green solvents for peptide manufacturing has yet to become routine in the industry. Given this reality, chemists continuously develop and seek incremental green chemistry strategies.

One major domain to exploit green chemistry concepts in peptide chemistry is to enhance peptide manufacturing productivity. The conventional SPPS strategy adopts stepwise assembly of the side-chain protected amino acid building blocks to solid supports, detaches the assembled peptide off the resin, and removes the side-chain protecting groups with a highly concentrated acid solution. The treated resin is removed, and the crude product is obtained by precipitating the peptide in a large volume of anti-solvent. Such a classic process is frequently restricted by small batch size and low productivity and is deemed contradictory to green chemistry concepts. An innovative MP-SPPS (Minimal-Protection Solid-Phase Peptide Synthesis) process is developed to address the root causes of the low productivity inherent to conventional SPPS. Radical productivity increase is accomplished on a peptide API manufacturing through the MP-SPPS process.

2. Low-manufacturing productivity of SPPS

One of the most evident peptide manufacturing issues entangled in adverse ecology effects is its relatively low productivity and high PMI (Product Mass Intensity). A large volume of organic solvents is generally utilized in SPPS, including those applied in the amino acid coupling, resin rinsing, peptide cleavage/global deprotection, and workup.

For example, small batch sizes of SPPS leading to poor productivity speaks for a common restriction in peptide industrial manufacturing, which frequently gives rise to the high costs of peptide API manufacturing. A QbD (Quality by Design) tool of the root-cause tree reveals that one of the prime reasons for the low productivity/high cost of peptide manufacturing originates from the amino acid side-chain protection (**Figure 1**).

Highly concentrated TFA solution is generally applied in the Fmoc-SPPS to cleave the peptides from the solid supports through acidolysis and remove the side-chain

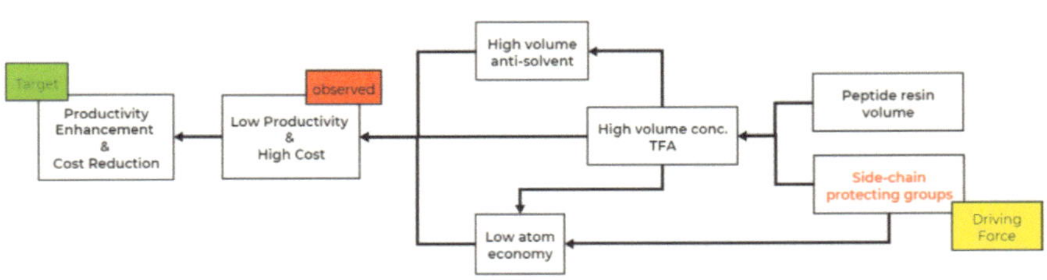

Figure 1.
Root-cause tree of the low productivity/high-cost analysis from the conventional SPPS.

protecting groups simultaneously. The physical volume of peptidyl resin and the presence of relatively stable protecting groups, for example, Pbf (2,2,4,6,7-pentamethyldihydrobenzofuran-5-sulfonyl) [8], often require 200 equivalent of TFA (sometimes much more) relative to the equivalent of peptide molecule. Industrial setting involving peptide manufacturing precludes the standard evaporation technologies, such as rotary and thin-film evaporation, of highly concentrated TFA mixture, due to the operability, safety, and stability restrictions. The large volume of the derived peptide TFA solution at the step of peptide cleavage/global deprotection thus poses a considerable challenge to the subsequent workup processes. It requests an even greater amount of anti-solvent, such as MTBE (methyl *tert*-butyl ether), to sufficiently precipitate the crude peptide product from its TFA solution. Generally, the efficient precipitation of the peptide product requisites 3–7 volumes of anti-solvent *cf*. peptide TFA solution. This setup poses an evident bottleneck to the whole process due to handling the extraordinarily large volume of solutions, and it pronouncedly reduces the productivity of the affected manufacturing, as revealed by **Figure 1**. Such demands pose a serious burden to the industries. Moreover, the TFA-labile side-chain protecting group's introduction-removal strategy adversely affects the atom economy of peptide synthesis.

Skipping the side-chain global deprotection step represents one of the potentially viable strategies to minimize the TFA quantity and the volume of the peptide solution prior to the product precipitation. It is not uncommon that the peptide cleavage/global deprotection step poses a bottleneck to the whole process, considerably lowering the batch size and limiting productivity. This artifact predominantly relates to the large volume of TFA applied at the step of peptide cleavage/global deprotection.

The acid-labile side-chain protecting groups requires a sufficiently high concentration of TFA in order to remove them from the masked functional groups quantitatively, and the physical volume of the peptidyl resin brings about the consumption of huge excess of TFA input (generally more than 200 equiv.) at the cleavage/global deprotection step. Incorporating side-chain unprotected residues in SPPS could significantly lower the TFA concentration required to detach the peptide molecules from the super acid-labile resins such as CTC, Sieber, and Ramage resins. Supplementing auxiliary low-boiling point solvents to the cleavage solution could decrease the TFA concentration and ease the following workup by stripping off the solvent under reduced pressure prior to the product precipitation. The resulting concentrated peptide mixture reduces the volume needed for the peptide precipitation step. Such a method improves peptide synthesis productivity and allows much greater batch sizes, which overall lowers manufacturing costs and aligns with green chemistry principles.

3. Minimal-protection SPPS (MP-SPPS)

3.1 Concept of MP-SPPS

Peptide synthesis generally requires the protection of the nucleophilic N^{α} and the side chain of the amino acid building block. The N^{α}-protecting groups, generally Fmoc in Fmoc/tBu SPPS strategy, are temporarily used and removed post the amino acid coupling. The side-chain protecting groups, on the contrary, remain anchored on the protected functional groups during the peptide solid-phase synthesis and are simultaneously removed during or post peptide cleavage [9].

Protecting groups shield the nucleophilic reactivity from the side chains' functional group on the peptide residues or the incoming reactive amino acid

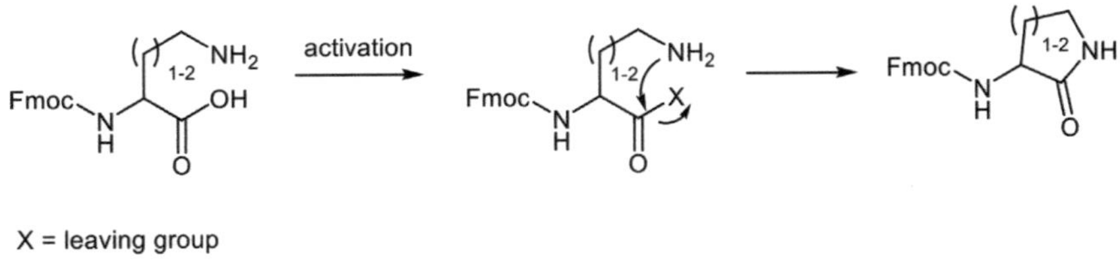

Figure 2.
Intramolecular cyclization of side-chain unprotected C^{γ}- and C^{δ}-substituted amino acid derivatives after carboxylate-activation.

intermediate. Such nucleophilic groups could react with the electrophilic coupling agent or the incoming reactive amino acid intermediate present during the peptide bond formation. Moreover, an intramolecular reaction could occur if the side-chain nucleophilic group remains unmasked. This undesired conversion readily affects the C^{γ}- and C^{δ}-substituted amino acid derivatives such as Dab (2,4-Diaminobutyric acid), Orn, and Arg (**Figure 2**).

Consequently, mandatory protection applies to N^{α}, Lys-N^{ε}, Orn-N^{δ}, Dab-N^{γ}, Dap (2,3-Diaminopropionic acid)-N^{β}, COOH$^{\beta}$-Asp, COOH$^{\gamma}$-Glu, S^{β}-Cys, and COOH$C^{\text{-terminal}}$ (in case of amino acid side-chain immobilization) to avoid undesired side reactions that predominantly occur during the peptide coupling step (**Figure 3**). The subject carboxylate, amino, and thiol groups from these amino acids constitute the reactive functional group in question. If not properly masked, the Fmoc-SPPS chemistry will challenge their integrities, yielding a complex mixture of by-products.

On the contrary, other relevant nucleophilic functional groups, even though traditionally protected in peptide synthesis, could survive the coupling reactions without protection under proper conditions. These tolerable functional groups encompass N^{δ},N^{ω},$N^{\omega'}$-Arg, the carboxamide group from Asn/Gln, O^{β}-Ser/Thr, the phenolic group from Tyr, N^{im}-His, and N^{indole}-Trp (**Figure 3**).

By taking advantage of such tolerable functional groups, the minimal-protection SPPS (MP-SPPS) strategy aims to utilize amino acid building blocks without side-chain protecting groups. In other words, MP-SPPS make use of Fmoc-Asn-OH, Fmoc-Gln-OH, Fmoc-Ser-OH, Fmoc-Thr, Fmoc-His-OH, Fmoc-Arg-OH, and Fmoc-Trp-OH without side-chain protection unless the occurrence of unforgiving and invincible side reactions.

Besides the aforementioned benefits related to the productivity boost, MP-SPPS enables the following improvements:

- Enabling considerably milder peptide cleavage conditions. Harsh conditions of peptide resin acid treatment, like an increased amount of TFA, the addition of strong acids such as trifluoromethanesulfonic acid, elevated temperature, and prolonged cleavage time might be wielded in case relatively stabler protecting groups could not be effectively removed under standard acidolysis conditions. In addition to representing a safer alternative, the implementation of MP-SPPS could preclude the harsh deprotection condition and lower the risk of acid-catalyzed side reactions such as Asn/Gln deamidation, aspartimide formation, and peptide backbone hydrolysis [10].

- Outrooting a plethora of underlying side reactions inherent to the protecting groups, such as incomplete protecting group removal, Pbf, in particular,

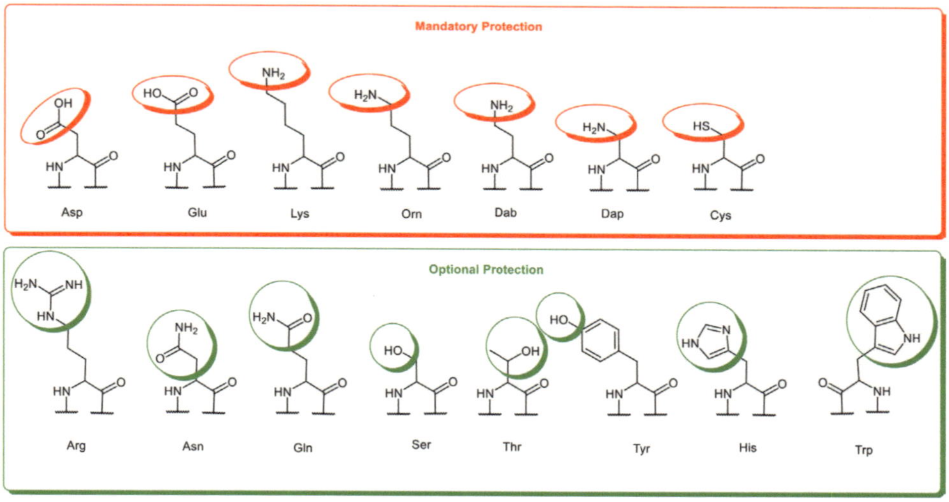

Figure 3.
Differentiation of the side-chain protection strategy in SPPS.

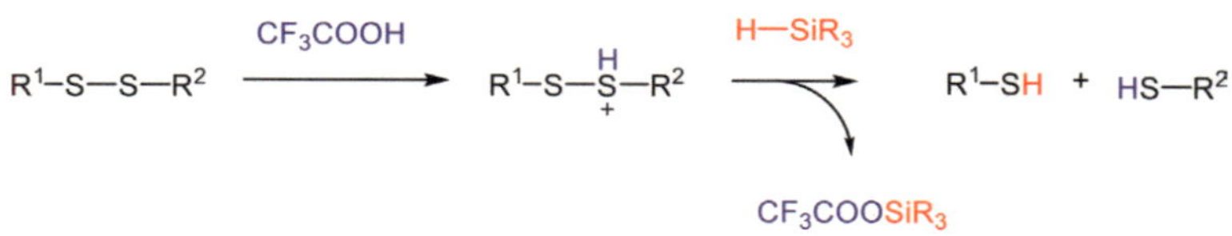

Figure 4.
TFA/Silane-induced disulfide reduction.

tert-butylation, sulfonation (formation of [M+80] impurities induced by Pbf degradation), trifluoroacetylation, and so on [11].

- Eliminating the scavengers, like silane, that entraps the released protecting group cations. The electrophilic scavengers could complicate the deprotection reaction and induce undesired side reactions such as disulfide breaking. Exclusion of the scavengers could therefore cement the stability of certain labile moieties, like disulfide bonds (**Figure 4**) [12].

- Improving the COGS (Cost of Goods Sold) by reducing TFA, anti-solvent, and scavenger, improving the atom economies and PMI.

- Lowering the negative ecological impact (E-factor enhancement).

- Strengthening manufacturing safety.

The concept of minimal-protection SPPS has been introduced previously. Strategies for assembling the target peptide products, including side-chain unprotected amino acids, have been reported hitherto, including ultrasound-facilitated peptide synthesis [13], amino acid couplings by restrainedly activated carboxylate species like *p*-nitrophenyl ester [14], or special building modalities such as iso-acyl dipeptide building block Boc-Ser/Thr(Fmoc-Xaa)-OH [15]. Side-chain unprotected amino acids could be utilized as the starting materials in these strategies for assembling the target peptide products. Nevertheless, the strategies apply mostly for small-scale syntheses or peptide derivatives with very short sequences, such as di- or tri-peptide. The development of MP-SPPS strategies fully compatible with large-scale industrial peptide production, including its applicability scopes, restrictions, and corresponding solutions, has yet to be fully elaborated. Overall, MP-SPPS subscribes to the growing interest in protecting group-free organic synthetic methodology [16].

In addition, functional groups, including N^{α}, Lys-N^{ε}, Orn-N^{δ}, Dab-N^{γ}, Dap-N^{β}, COOH$^{\beta}$-Asp, COOH$^{\gamma}$-Glu, and S^{β}-Cys, still require proper protection under MP-SPPS.

3.2 Side-chain unprotected arginine

Arg stands out as the most influential residue for the MP-SPPS among the non-mandatory side-chain protection amino acid species. Arg's default side-chain protecting group Pbf is notoriously known for its acidolytic stability. A peptide of interest rich in Arg residues exacerbates the problem. Such situations call for the use of a high concentration of TFA. Furthermore, auxiliary stronger acid such as TFMSA (trifluoromethanesulfonic acid) is occasionally needed to boost the Pbf cleavage. Given the challenges inherent to Pbf acidolysis, MP-SPPS should target incorporating side-chain unprotected Fmoc-Arg-OH as a prime task.

The extremely low conversion rate of Fmoc-Arg-OH coupling impedes its utilization as a side-chain unprotected building block when assembled to a peptidyl resin Xaa^{3}-Tyr-Xaa^{5}-His-Xaa^{7}-Xaa^{8}-2-chlorotrityl resin (Xaa^{3,5,7,8} are unreactive aliphatic or aromatic amino acids without nucleophilic side chains). In this case, the use of 1.5 equiv. Fmoc-Arg-OH with 1.5 equiv. DIC (*N*,*N'*-diisopropylcarbodiimide)/HOSu (*N*-hydroxysuccinimide) in DMF does not lead to the desired product formation even after 3 h of reaction. Alternative couplings additives Oxyma (Ethyl cyano(hydroxyimino)acetate) or HOBt (*N*-hydroxybenzotriazole) in DMF or NMP do not facilitate the subject coupling. Furthermore, various combinations of coupling additives and organic solvents could not form the desired peptide product. RP-HPLC analysis of the Fmoc-Arg-OH/DIC/Oxyma coupling solution reveals a rather complex mixture (**Figure 5**).

LC/MS reveals the cyclization and dimerization of the Fmoc-Arg-OH upon the carboxyl activation. Concretely speaking, Fmoc-Arg-OH (1) transforms to Fmoc-Arg-lactam (2), Fmoc-Arg(Fmoc-Arg)-OH (3), and the dimeric lactam (4) (**Figures 5** and **6**).

Apparently, the undesired side reactions addressing Fmoc-Arg-OH prior to its condensation with the peptide N^{α}-group explain the low coupling yield. The guanidino functional groups are stereotypically deemed inert under common reaction conditions. However, the guanidino group of arginine, equating a C^{δ}-derivatized ornithine, enables the intramolecular cyclization depicted in **Figures 2** and **6**. The side-chain unprotected Fmoc-Arg-OH undergoes a kinetically favored six-membered ring cyclization with its activated carboxylate, yielding an unreactive lactam. The arginine guanidino group is not as inert as generally believed. Actually, freebase guanidine compounds constitute strong nucleophilic species and are capable of nucleophilic additions [17]. Arginine is generally tagged as unreactive simply because its guanidino side chains are mostly protonated under acidic to mild basic conditions thanks to its high pK_a value (>12). Nonetheless, when the guanidino group is partially deprotonated, it can initiate a nucleophilic attack on the proper electrophilic substrates, as

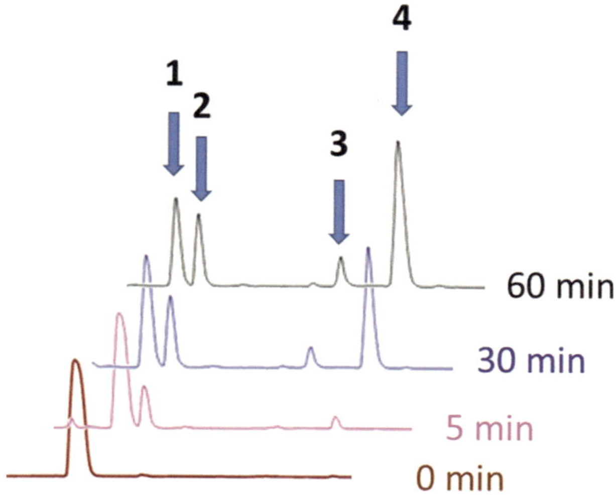

Figure 5.
RP-HPLC traces of the solutions of Fmoc-Arg-OH activation by DIC/Oxyma in DMF [1. Fmoc-Arg-OH; 2. Fmoc-Arg-lactam; 3. Fmoc-Arg(Fmoc-Arg)-OH; 4. Fmoc-Arg(Fmoc-Arg)-lactam].

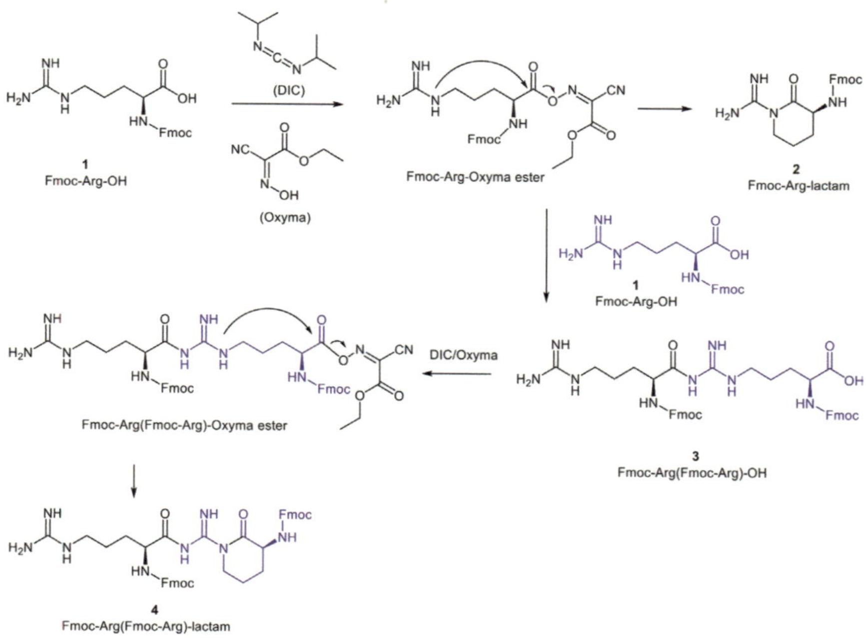

Figure 6.
Fmoc-Arg-OH transformation by DIC/Oxyma treatment.

indicated in **Figure 6**. Therefore, Fmoc-Arg(Pbf)-OH is commonly used in Fmoc-SPPS. The bulky protecting group Pbf minimizes the risk of the aforementioned side reaction.

Nevertheless, the inherent issues centered on Fmoc-Arg-OH could be circumvented by using Fmoc-Arg(HCl)-OH, which undermines the Fmoc-Arg-OH unprotected guanidino side's reactivity. The salt formation of Fmoc-Arg-OH simultaneously protonates the nucleophilic guanidino group and introduces steric interference by the chloride counterion. Furthermore, using a carbodiimide-based coupling agent does not require the utilization of an auxiliary base catalyst. With the undesired intramolecular cyclization minimized, the target amino acid coupling is cemented (**Figure 7**). Indeed, quantitative coupling of 1.5 equiv. Fmoc-Arg(HCl)-OH is accomplished by DIC/Oxyma in DMF, validating the utility of Fmoc-Arg(HCl)-OH as arginine synthon for the MP-SPPS strategy.

3.3 Side-chain unprotected histidine

The side-chain unprotected histidine Fmoc-His-OH undergoes considerable imidazolyl modification by DIC during MP-SPPS, thanks to the pronounce nucleophilicity of imidazolyl group toward carbodiimide, resulting in an [M + 126] impurity (**Figure 8**).

Unlike the arginine cyclization side reaction delineated in **Figure 6**, the DIC-induced imidazolyl modification is progressive. It intensifies through the MP-SPPS process once the subject histidine residue is incorporated into the immobilized peptide chain. This artifact precludes utilizing Fmoc-His-OH as the building block when carbodiimide-based coupling reagents orchestrate the MP-SPPS processes.

Phosphonium salt coupling reagents such as PyBOP do not modify the histidine imidazolyl side chain under work conditions during the peptide assembly. This merit cements the intrinsic advantages of phosphonium coupling reagents which are inert to the N^{α} functional group during peptide synthesis.

Incorporating side-chain unprotected Fmoc-Arg(HCl)-OH and Fmoc-His-OH could proceed well when the histidine is on the *N*-flank of the arginine. The assembly of Fmoc-Arg(HCl)-OH and the subsequent residues could be mediated by DIC/Oxyma strategy prior to the Fmoc-His-OH incorporation (**Figure 9a**). Then, the

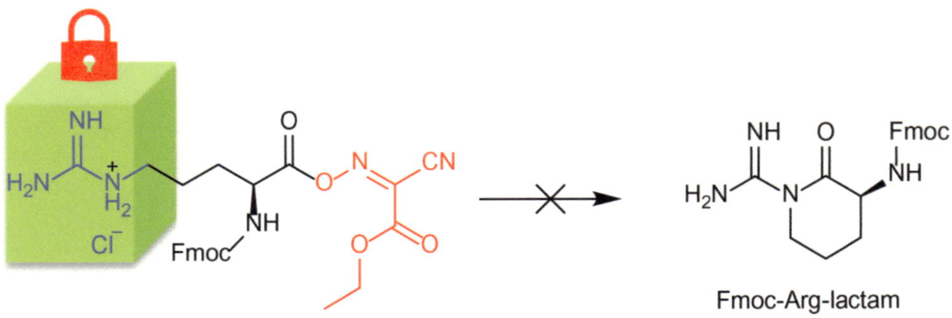

Figure 7.
Strategy to suppress the undesired lactam formation during Fmoc-Arg-OH carboxylate activation.

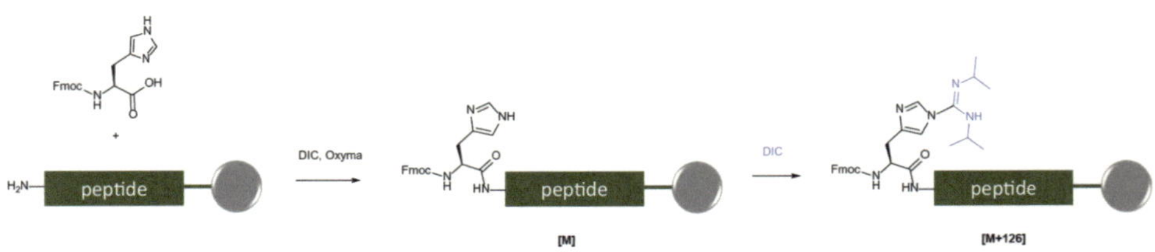

Figure 8.
Modification of N^{im}*-His by DIC.*

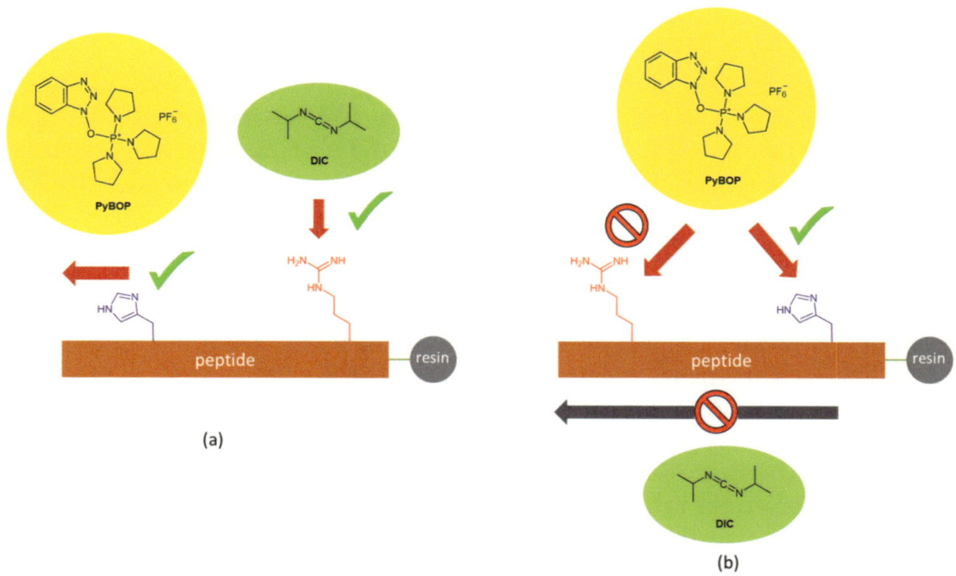

Figure 9.
Coupling reagent suitability for various Arg/His topologies: (a) the unprotected His is N-flank to the unprotected Arg; (b) the unprotected His is C-flank to the unprotected Arg.

coupling reagents mixture uses PyBOP/base combination for all the following amino acid couplings. Otherwise, DIC will modify the imidazolyl side chain of the histidine residue already incorporated into the peptide chain attached to the solid support. On the contrary, an arginine *N*-terminal to the histidine, not necessarily in direct proximity, creates a dilemma for the MP-SPPS due to the reactivity of the histidine imidazolyl group toward carbodiimide coupling reagents and the requirement of the utilization of phosphonium salt to mediate all the amino acid couplings *N*-flanking to the subject histidine (**Figure 9b**). In this scenario, PyBOP will constitute the coupling reagent of choice to mediate all the amino acid couplings from the Fmoc-His-OH onwards, including the Fmoc-Arg(HCl)-OH. However, incomplete Fmoc-Arg(HCl)-OH incorporation might arise as a consequence. PyBOP is incapable of driving the Fmoc-Arg(HCl)-OH coupling into completion in the subject peptide assembly.

The incorporation of a side-chain unprotected Fmoc-Arg(HCl)-OH in a -Arg-$(Xaa)_n$-His- peptide topology will necessitate the usage of an alternative histidine synthon to satisfy the MS-SPPS demand. Because the major positive impact of MP-SPPS depends predominantly on reducing the amount and concentration of TFA during side-chain deprotection, the utilization of side-chain unprotected Arg should be prioritized. In other words, the reconciliation is made in favor of Fmoc-Arg(HCl)-OH relative to Fmoc-His-OH. In this scenario, one could resort to more acid-sensitive protecting groups for histidine, such as Mtt (methyltrityl) or Mmt (monomethoxytrityl). The acid-lability is ranked in the following incremental order of Trt < Mtt < Mmt. Moreover, a case study reveals that Fmoc-His(Mtt)-OH could be readily incorporated into the immobilized peptide chain by DIC/Oxyma chemistry with reduced racemization relative to that of Fmoc-His(Trt)-OH during peptide coupling [18]. More importantly, 10% TFA/DCM with 1.2 equiv. TIS (triisopropylsilane), as the Mtt^+ scavenger, removes the Mtt-protecting group effectively and sufficiently.

3.4 Side-chain unprotected serine/threonine

The nucleophilicity of the hydroxyl group-bearing amino acids couplings, like Fmoc-Ser-OH, Fmoc-Thr-OH, and Fmoc-Tyr-OH, represents an inherent challenge for MP-SPPS. Fmoc-Ser-OH, as a C^{β}-hydroxy substituted amino acid, could mediate a double insert of serine residue and the resultant formation of endo-Ser impurity. The process is depicted in **Figure 10**.

A second Fmoc-Ser-OH could add to the unprotected hydroxyl group of another serine, either post or prior to the immobilization on the solid phase but during the coupling reaction. After removing the Fmoc protecting group from the Fmoc-Ser(Fmoc-Ser)-OH residue of the peptide chain attached to the solid support, the side-chain serinyl moiety could migrate to the peptide backbone through a favored five-membered ring intermediate initiated by the N^{α}-group (**Figure 10**). This acyl $O \rightarrow N$ shift is particularly facilitated in the basic milieu and results in the endo-Ser impurity formation with a molecular weight increase of 87 Da. Without the acyl $O \rightarrow N$ shift, the residual serinyl ester bond will be hydrolyzed at the subsequent piperidine treatment. The liberated hydroxyl group might further mediate ester bond formation at the subsequent amino acid coupling step. However, the formed ester impurity will unlikely lead to an acyl $O \rightarrow N$ shift due to the steric disadvantages of the greater ring size, which precludes the insertion of an extra residue in the peptide sequence. Fmoc-Thr-OH could, in principle, also accommodate this process, but the extent is much restrained due to the steric hindrance of the C^{β}-methyl group.

In order to address the undesired serine side-chain esterification, process parameters such as amino acid pre-activation temperature, pre-activation time, coupling temperature, pH, solvent, coupling reagent, and coupling additive could be screened and tuned to minimize the endo-Ser impurity formation.

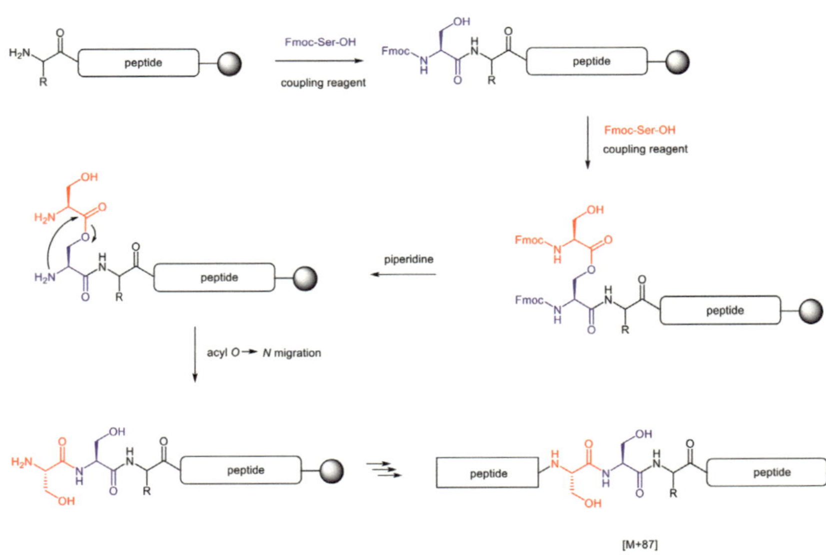

Figure 10.
Formation of endo-Ser impurity from Fmoc-Ser-OH coupling.

3.5 Side-chain unprotected tyrosine

The phenolic group from the tyrosine side chain could also be subjected to phenyl ester formation in the MP-SPPS process but to a much-reduced extent due to the weaker nucleophilicity of phenolic hydroxyl group relative to the aliphatic hydroxyl group on serine (**Figure 11**). The undesired function between tyrosine's phenolic group with the coupling reagent has also been reported [19].

Notably, the susceptibility of tyrosine to esterification might be sequence dependent. In this study, the Tyr(Fmoc-Xaa^n) impurity formation occurs exclusively at a specific Fmoc-Xaa^n-OH coupling step without being detected at other coupling steps [17]. Spatial alignment might come into play to facilitate the tyrosine-induced esterification through acyl $N \rightarrow O^{phenolic}$ shift. Meanwhile, with this undesirable transformation taking place, the excess Fmoc-Xaa^n-OH in the reaction solution can acylate the liberated N^{α}-group to yield a derivative of [Tyr(Fmoc-Xaa^n)] peptide intermediate (**Figure 11**). The piperidine treatment removes the N^{α}-Fmoc protecting group and the Xaa^n ester from the tyrosine side chain, regenerating the targeted peptide sequence. Side-chain phenyl ester is facilely hydrolyzed in this process, and the target intermediate is subject to the following SPPS steps to yield the target peptide product. This process is schematized in **Figure 11.**

3.6 Side-chain unprotected asparagine/glutamine

Side-chain unprotected Fmoc-Asn/Gln-OH also sustains the MP-SPPS. The most prevalent side reaction arising during the Asn activation and assembly is the dehydration and the resulting formation of a β-cyano side chain. The process involves the

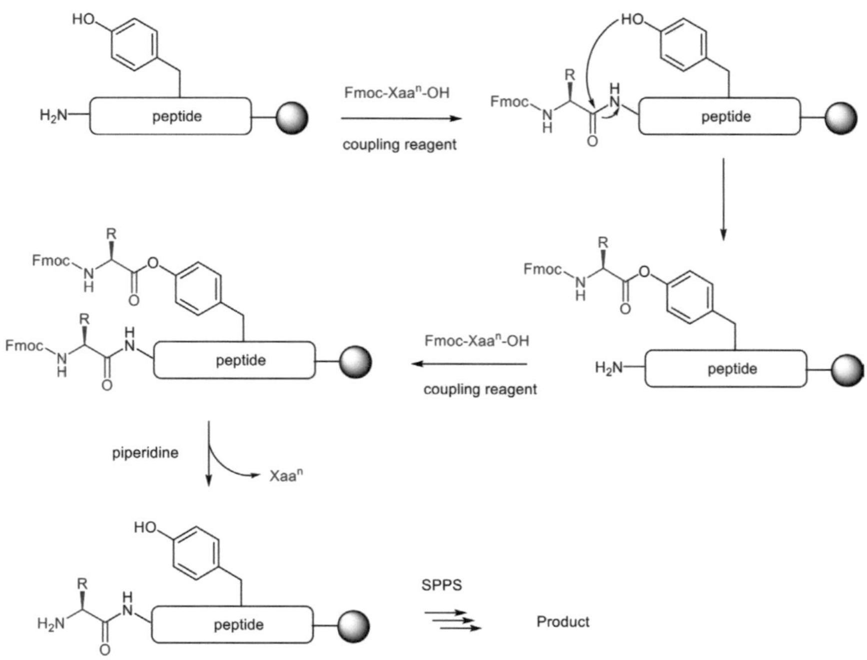

Figure 11.
Putative mechanism of tyrosine esterification in MP-SPPS.

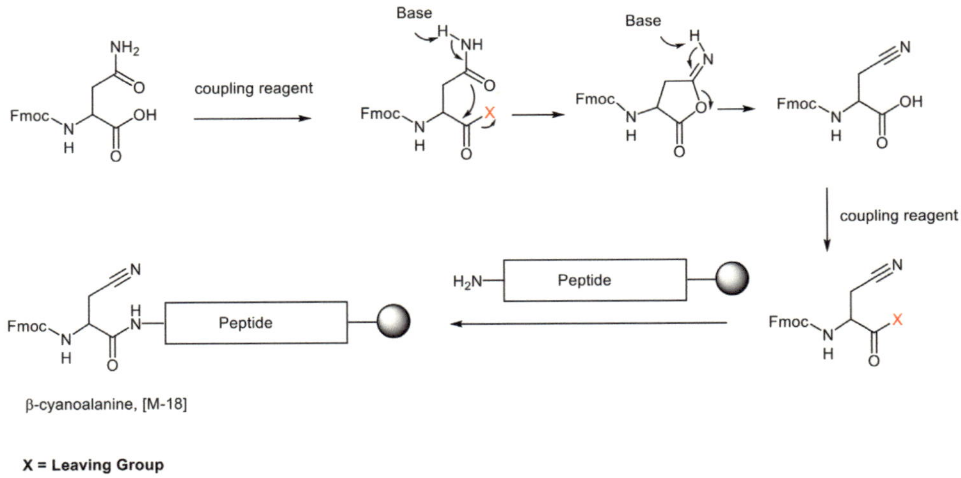

Figure 12.
Formation of β-cyanoalanine impurity through side-chain unprotected Asn activation and dehydration.

unprotected amide side chain attacking the activated carboxylate in the presence of a base, resulting in a five-membered ring intermediate iminohydrofuranone. The latter undergoes ring opening in basic conditions and gives rise to the formation of Fmoc-Ala(CN)-OH (**Figure 12**) [20]. The coupling reagent present in the reactive mixture can activate the resulting Fmoc-Ala(CN)-OH by-product, which incorporates into the peptide chain instead of the desired Fmoc-Asn-OH and yield an [M-18] impurity.

This undesired conversion occurs only at the Fmoc-Asn-OH activation step when incorporated into the growing peptide chain. No further dehydration will process once the side-chain unprotected Asn is part of the peptide chain. The activation reaction and the existence period of the activated Fmoc-Asn-OH species dictate the extent of the Fmoc-Asn-OH dehydration. Process parameters, including coupling reagent, additive, solvent, amino acid activation pattern, pre-activation temperature, pre-activation time, and coupling temperature, might impact the propensity of the Fmoc-Asn-OH dehydration.

4. A case study of MP-SPPS

The incorporation of side-chain unprotected Arg, Tyr, and Fmoc-His(Mtt)-OH during the SPPS on CTC resin of an octa-peptide of angiotensin II analog called peptide Z has been reported [17]. Peptide Z bears a sequence of H-Xaa^1-Arg-Xaa^3-Tyr-Xaa^5-His-Xaa^7-Xaa^8-OH (Xaa^{1,3,5,7,8} are unreactive aliphatic or aromatic amino acids without nucleophilic side chains). The couplings of 1.5 equiv. Fmoc-Arg(HCl)-OH, Fmoc-His(Mtt)-OH, and Fmoc-Tyr-OH by DIC/Oxyma in DMF go quickly and quantitatively without forming any detectable related impurities. Interestingly, completing Fmoc-Arg(HCl)-OH coupling takes 2.5 h, as attested by a negative ninhydrin test, whereas all the other couplings are entirely completed within 1.5 h. The MP-SPPS strategy achieves quantitative couplings without the occurrence of major side reactions/incomplete coupling.

The peptide cleavage is pivotal to the MP-SPPS strategy as the following criteria should be gratified:

- The acid cleavage solution should quantitatively release the immobilized peptide molecules from the solid supports.
- The acid treatment should quantitatively deprotect the acid-labile side-chain protecting groups on the peptide.
- The cleavage and deprotection solution should require the lowest possible acid concentration to accomplish the quantitative reactions.
- The method should easily concentrate the peptide solution post the cleavage and deprotection without compromising the product integrity.
- The peptide solution should be miscible with the anti-solvent at the product workup step; otherwise, solvent swapping should be performed before the product precipitation.
- The solvent for the acid cleavage solution should comply with the industrial request. The low OEL (Occupational Exposure Limit) and high toxicity (confirmed animal carcinogen and probable carcinogen to humans) of DCM disqualify its utilization [21].

To address the MP-SPPS demands during peptide release from the solid support and deprotection, 10% TFA/DCM with 1.2 equiv. TIS (as the scavenger for the Mtt cation) could accomplish the above criteria except for the use of DCM. As a result, peptide Z is fully cleaved from the 2-chlorotrityl resin, and the Mtt protecting group on the His side chain is quantitatively removed simultaneously after 1 h at the ambient temperature.

To comply with all the MP-SPPS criteria, a solvent screen has been conducted to find the replacement for DCM. Among ACN, 2-MeTHF, IPA, EtOH, EtOAc, and TFT (trifluorotoluene), only TFT enables the successful replacement of DCM as the bulky solvent in conjunction with TFA for the one-pot peptide deprotection and release from the solid support.

Various reactions such as acylation, tosylation, silylation, Swern oxidation, and Dess-Martin oxidation have been accomplished by replacing DCM with TFT as the organic solvent [22]. TFT has a much higher boiling point (102°C) than DCM (40°C), granting improved safety profiles relative to DCM in the industries. Consequently, 10% TFA/TFT/1.2 equivalents of TIS (17 equivalents of TFA *cf*. peptide) at room temperature for 2 h gives a quantitative peptide Z cleavage and Mtt removal. Notably, the hydrophilic peptide Z could not fully solubilize in the relatively unpolar 10%TFA/TFT/TIS cleaving/deprotecting solution, even though the solution enables the quantitative cleavage of the peptide Z from the resin. Nonetheless, rinsing the cleaved resin three times with 10% HFIP (hexafluoroisopropanol)/TFT drastically enhanced the product recovery by effectively solubilizing the peptide Z in HFIP.

Interestingly, the peptide Z TFA/HFIP/TFT solution undergoes phase separation prior to the concentration step. The TFT phase contains nearly no product but only the by-product methyl triphenylmethane derived from Mtt cation quenched by TIS. The salting-out effect might explain the phase separation of the inherently miscible TFT and HFIP [23], which has been exploited by SALLE (salting-out assisted liquid-liquid extraction) for sample analysis [24]. The serendipitous phase separation affords a viable methodology to isolate peptide Z through tandem procedures of phase separation, facile condensation of the peptide HFIP solution (optional), and product precipitation. Despite such potential imparted by the phase separation, the combined peptide cleavage /rinsing solution in this study is subjected to solvent swapping from TFT to HFIP. No product degradation is detected in this process. The peptide Z precipitates with cold MTBE once added to the concentrated peptide/HFIP mixture to obtain a quantitative uncorrected total yield of SPPS/cleavage/isolation.

The LC/MS analysis reveals a crude product purity of 91.7%, comparable to those from various industrial batches (90–94%).

Figure 13 and **Table 1** compare the conventional SPPS and MP-SPPS in terms of process and performance index, respectively.

The most striking attributes inherent to MP-SPPS are the radically reduced TFA consumption (by 95%) and the increased peptide concentration prior to the precipitation (by 4.3 folds). These two key indices enable appreciable increases in peptide manufacturing productivity through MP-SPPS. The overall precipitation volume (combination of peptide solution and the anti-solvent) enhances productivity 5.3-fold when taken as a baseline since it poses the bottleneck to the whole manufacturing

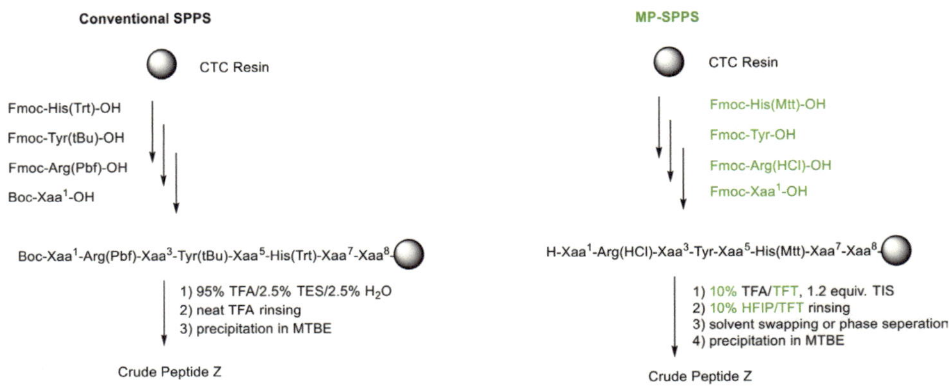

Figure 13.
Comparison of peptide Z synthetic scheme from conventional SPPS and MP-SPPS.

Index	SPPS	MP-SPPS
Atom economy (amino acid)	22.3%	23.9%
Overall TFA (cleavage + rinsing)	25.6 L/mol	1.3 L/mol
TFT	0	23.9 L/mol
HFIP	0	7.8 L/mol
Concentration prior precipitation	0.038 M	0.2 M
MTBE as anti-solvent	78.9 L/mol (3 vol.)	15 L/mol (3 vol.)
Total volume (sum of MTBE and Peptide solution) of precipitation	105 L/mol (3 vol. MTBE)	20 L/mol (3 vol. MTBE)
Crude purity	90–94%	92%
Uncorrected yield (crude product)	quantitative	quantitative

MP-SPPS could have a 5.3-fold of the conventional SPPS productivity.

Table 1.
Comparison of the major indices between conventional SPPS and MP-SPPS for peptide Z synthesis.

process (**Table 1**). Overall, the quality and yield of the crude peptide Z produced from MP-SPPS are comparable to conventional SPPS.

5. Summary

The MP-SPPS strategy reduces the hazardous elements of SPPS. It represents an environmentally friendlier method and enhances manufacturing productivity when using the side-chain unprotected amino acids such as Arg, His, Ser, Thr, Tyr, Asn, and Gln. Because the targeted peptide sequence may affect the Fmoc-Arg(HCl)-OH coupling efficacity, thoughtful planning should include a pilot study. Nevertheless, excluding the Pbf protecting group alleviates the global deprotection processes appreciably, which generally request highly concentrated TFA, occasional supplementation of stronger acids, and prolonged reaction time. Furthermore, lowering the TFA concentration allows for recovering a peptide solution with a higher concentration, which in turn eases the peptide isolation by precipitation.

Consequently, this procedure enhances the productivity of peptide manufacturing and fully outroots the frequent Pbf-related sulfonation impurity [M+80]. In addition, MP-SPPS lowers the occurrence of highly concentrated TFA-induced impurities such as trifluoroacetylation [M+96], backbone hydrolytic cleavage, acid-induced aspartimide formation [M-17/18], acid-catalyzed disulfide bond reduction by silane [M+2] [9], and deamidation of carboxamide moieties [M + 1]. Deamidation impurity represents a serious general concern as chromatographic purification techniques frequently fail to remove them from the desired peptide products.

Notably, the efficacy of Fmoc-Arg(HCl)-OH coupling might be sequence-dependent. The MP-SPPS strategy does not provide a solution to the use of indispensable side-chain protection (such as Lys, Asp, and Glu) if the targeted peptide sequence includes such residues. However, adopting alternative orthogonal protecting groups could diminish the highly concentrated TFA mixture application. For instance, Pd^0-catalyzed allyl ester cleavage removes the allyl ester protecting group from a carboxylate side chain of Asp and Glu.

In summary, the MP-SPPS process addresses the high demand for developing greener chemical processes and lowering peptide manufacturing costs. MP-SPPS aligns with important Green Chemistry Principles [25], including atom economy (improved PMI), less hazardous chemical syntheses (TFA), safer solvents/auxiliaries (TFT vs. DCM), design for energy efficiency (skipping peptide solution concentration through phase separation), reduce derivatives (elimination of the unnecessary protecting group introduction-elimination), and inherently safer chemistry for accident prevention (no TFA evaporation, no handling of bulky highly concentrated TFA solution). Attainable superiorities in productivity, cost-effectiveness, and E-factor (environmental factor) constitute the three pivotal pillars for applying the MP-SPPS in industries.

Acknowledgements

The author would like to thank Antoine Blanc, Ph.D., from The State University of New York College at Oneonta, for reviewing the manuscript.

Author details

Yi Yang
Chemical Development, Global Pharmaceutical R&D, Ferring Pharmaceuticals A/S, Denmark

*Address all correspondence to: yi.yang@ferring.com

References

[1] Jadhav S, Martin V, Egelund PHG, Castro HJ, Krüger T, Richner F, et al. Replacing DMF in solid-phase peptide synthesis: Varying the composition of green binary solvent mixtures as a tool to mitigate common side-reactions. Green Chemistry. 2021;**23**:3312-3321

[2] Ferrazzano L, Catani M, Cavazzini A, Martelli G, Corbisiero D, Cantelmi P, et al. Sustainability in peptide chemistry: Current synthesis and purification technologies and future challenges. Green Chemistry. 2022;**24**:975-1020

[3] Wegner K, Barnes D, Manzor K, Jardine A, Moran D. Evaluation of greener solvents for solid-phase peptide synthesis. Green Chemistry Letters and Reviews. 2020;**14**:153-164

[4] Ferrazzano L, Corbisiero D, Martelli G, Tolomelli A, Viola A, Ricci A, et al. Green solvent mixtures for solid-phase peptide synthesis: A dimethylformamide-free highly efficient synthesis of pharmaceutical-grade peptides. ACS Sustainable Chemistry & Engineering. 2019;7:12867-12877

[5] Martin V, Egelund PHG, Johansson H, Le Quement ST, Wojcik F, Pedersen DS. Greening the synthesis of peptide therapeutics: an industrial perspective. RSC Advances. 2020;**10**:42457-42492

[6] Jaradat DMM, Al Musaimi O, Albericio F. Advances in solid-phase peptide synthesis in aqueous media (ASPPS). Green Chemistry. 2022;**24**:6360-6372

[7] Al Musaimi O, de la Torre BG, Albericio F. Greening Fmoc/tBu solid-phase peptide synthesis. Green Chemistry. 2020;**22**:996-1018

[8] Carpino LA, Shroff H, Triolo SA, Mansour E-SME, Wenschuh H, Albericio F. The 2,2,4,6,7-pentamethyldihydrobenzofuran-5-sulfonyl group (Pbf) as arginine side chain protectant. Tetrahedron Letters. 1993;**34**:7829-7832

[9] Isidro-Llobet A, Álvarez M, Albericio F. Amino acid-protecting groups. Chemical Reviews. 2009;**109**:2455-2504

[10] Yang Y. Peptide fragmentation/deletion side reactions, intramolecular cyclization side reactions. In: Side Reactions in Peptide Synthesis. Academic Press; 2015. pp. 1-28, 119-156

[11] Yang Y. Peptide global deprotection/scavenger-induced side reactions. In: Side Reactions in Peptide Synthesis. Academic Press; 2015. pp. 63-67

[12] Yang Y, Hansen L, Badalassi F. Investigation of on-resin disulfide formation for large-scale manufacturing of cyclic peptides: A case study. Organic Process Research and Development. 2020;**24**:1281-1293

[13] Anuradha MV, Ravindranath B. Acylation of unprotected amino acids using ultrasound. Tetrahedron. 1997;**53**:1123-1130

[14] Gagnon P, Huang X, Therrien E, Keillor JW. Peptide coupling of unprotected amino acids through in situ p-nitrophenyl ester formation. Tetrahedron Letters. 2002;**43**:7717-7719

[15] Arenzo HB, Bingenheimer W, Blanchette R, Morgans JRDJ, Robinson J III. Temporary serine protection in solid phase synthesis of LH-RH analogs.

International Journal of Peptide Protein Research. 1993;**41**:342-346

[16] Fernandes RA, editor. Protecting-Group-Free Organic Synthesis: Improving Economy and Efficiency. Wiley; 2018

[17] Taylor JE, Bulla SD, Williams JMJ. Amidines, isothioureas, and guanidines as nucleophilic catalysts. Chemical Society Reviews. 2012;**41**:2109-2121

[18] Yang Y, Hansen L, Ryberg P. Side-chain unprotected Fmoc-Arg/His/Tyr-OH couplings and their application in solid-phase peptide synthesis through a minimal-protection/green chemistry strategy. Organic Process Research and Development. 2022;**26**:1520-1530

[19] Vrettos EI, Sayyad N, Mavrogiannaki EM, Stylos E, Kostagianni AD, Papas S, et al. Unveiling and tackling guanidinium peptide coupling reagent side reactions toward the development of peptide-drug conjugates. RSC Advances. 2017;7:50519-50526

[20] Stroup AN, Cole LB, Dhingra MM, Gierasch LM. Synthesis and crystal structures of Boc-L-Asn-L-Pro-OBzl·CH_3OH and dehydration side product, Boc-β-cyano-L-alanine-L-Pro-OBzl. International Journal of Peptide and Protein Research. 1990;**36**:531-537

[21] Agency for Toxic Substances and Disease Registry Division of Toxicology and Environmental Medicine. Addendum to the Toxicological Profile for Methylene Chloride. Atlanta; 2010

[22] Ogawa A, Curran DP. Benzotrifluoride: A useful alternative solvent for organic reactions currently conducted in dichloromethane and related solvents. The Journal of Organic Chemistry. 1997;**62**:450-451

[23] Hyde AM, Zultanski SL, Waldman JH, Zhong Y-L, Shevlin M, Peng F. General principles and strategies for salting-out informed by the hofmeister series. Organic Process Research and Development. 2017;**21**:1355-1370

[24] Tang YQ, Weng N. Salting-out assisted liquid–liquid extraction for bioanalysis. Bioanalysis. 2013;**5**:1583-1598

[25] Anastas PT, Warner JC. Green Chemistry: Theory and Practice. New York: Oxford University Press; 1998. p. 30

Chapter 9

Renewable Energy and Green Technology

Suvojit Maity, Soham Sarkar, Kakoli Dutta, Tina De and Ruchira Mukherjee

Abstract

Our present need is to reduce or minimize the dependency on fossil fuels through a thorough understanding of the principles and utility of renewable energy along with green technology. It has been considered that renewable energy as an alternate source is the best and cheapest source of energy. The unwarranted usage of fossil fuels has a huge effect on climate change globally. This can be restored by the efficient utilization of renewable energy resources. Efficient utilization of renewable energy resources could minimize the impact of climate change globally. Generally, renewable energy is generated from essentially inexhaustible sources, including wind power, solar power, geothermal energy, tidal energy, biomass energy, and other sources. Sustainability researchers have highlighted the importance of green energy and its immediate impact on the environment. Therefore, transitioning toward green energy is crucial to reducing the global crisis and achieving sustainable development.

Keywords: green energy, sustainability, fossil fuels, renewable resources, green technology

1. Introduction

Since early ages, the Earth has depended on various forms of energy for its survival. From the initial age of the Industrial Revolution, fossil fuels have been used to generate energy for the needs of our daily lives. The first use of fossil fuels for energy production dates back to the early 1800s when coal was used to power steam engines. As the demand for energy increased globally during the Industrial Revolution, there was a need to create an alternative option besides fossil fuel to generate electricity and this was followed by the development of oil and natural gas as sources of energy in the late 1800s and 1900s.

The field of green chemistry has been attracting scientists and researchers for decades. Green chemistry is an innovative technology that reduces environmental damage and waste generated during chemical processes. It has introduced new terms such as "eco-efficiency," "sustainable chemistry," "atom efficiency," "process intensification and integration," "inherent safety," "product life cycle analysis," "ionic liquids," "alternate feedstocks," and "renewable energy sources." Its sixth principle offers a new approach for modifying known synthetic reactions in a sustainable manner. This involves replacing organic solvents with non-organic media to eliminate the volatility and corrosiveness of hazardous solvents, which helps to preserve the environment. The 12 principles of green chemistry are widely accepted criteria for comparing the environmental acceptability of these two processes. However, these principles do not fully explain the concept of green chemistry, which includes monitoring the lifecycle of fundamental processes and products, as well as recovering heat from exothermic or endothermic reactions [1].

In recent years, the use of fossil fuels such as oil and natural gas for energy production has increased dramatically, leading to widespread dependence on non-renewable energy sources. Unfortunately, this trend has also resulted in significant environmental consequences. To combat this, a transition toward low-carbon solutions is essential to reduce energy-related carbon dioxide (CO2) emissions, which make up two-thirds of all greenhouse gases (GHG). This transition can be made possible through technological innovation, particularly in the field of renewable energy. Thanks to rapidly falling costs and competitiveness, renewable energy sources such as solar photovoltaics (PV) and wind power have seen record new installations. One-quarter of all electricity worldwide was produced from renewables in the year 2017. However, the transition is not happening fast enough: After 3 years of steady CO_2 energy emissions from 2014 to 2016, they rose by 1.4% in 2017 [1].

Nowadays, about 45% of the total amount of carbon dioxide in the air comes from the burning of fossil fuels. It also releases other greenhouse gases like methane and water vapor which trap the heat coming from the Sun leading to "global warming." The burning of fossil fuels also releases a range of harmful pollutants, such as sulfur dioxide (SO_2), nitrogen oxides (NOx), and particulate matter into the air. This leads to air pollution, which has been linked to respiratory and other idiopathic diseases. Fossil fuels are finite resources and their continued use without alternative energy sources will lead to their depletion. This could lead to energy shortages, price spikes, and geopolitical tensions as countries compete for limited resources [2].

2. Concept of green energy

The word "renewable" means, capable of being renewed. Renewable energy is the type of energy that is produced from sources that are naturally replenishing and do not deplete over time such as solar energy, wind energy, hydroenergy, and so forth. Green energy is that which comes from natural sources, such as the sun, wind, water, tides, geothermal heat, or algae.

Though there are some differences between green energy and renewable energy, in most cases, they are considered just the same. In all likelihood solar energy, wind energy, hydrothermal energy, geothermal energy, vibrational energy, and biomass energy all are green as well as renewable. The concept of green energy is based on the principles of sustainable development, which aim to meet the needs of the present generation without compromising the ability of future generations to meet their own needs.

Green energy has become increasingly popular in recent years as more individuals, businesses, and governments recognize the importance of sustainability and environmental responsibility. Many countries have set ambitious targets for renewable energy adoption, and some have even committed to reaching net-zero emissions in the coming decades. While there are still challenges to overcome in the adoption of green energy, such as the need for energy storage solutions and infrastructure upgrades, the benefits are clear. By transitioning to renewable energy sources, we can create a more sustainable and resilient energy system that benefits both current and future generations [3, 4].

3. The need for green energy

Green energy is becoming increasingly important as the world becomes more aware of the negative impact of fossil fuels on the environment and the need to reduce carbon emissions. One of the key benefits of green energy is its ability to reduce greenhouse gas emissions. Burning fossil fuels contributes to both global warming and climate change. Renewable energy sources such as wind, solar, and hydropower produce little or no emissions, helping to mitigate the effects of climate change. Green energy is also a key driver of economic growth and job creation. The renewable energy industry employs millions of people around the world, from engineers and scientists to construction workers and installers. As countries invest in renewable energy, they are also creating new markets for products and services related to clean energy, which can boost local economies.

Another important benefit of green energy is its ability to increase energy security. Countries that rely heavily on imported fossil fuels are vulnerable to price fluctuations and supply disruptions. In contrast, renewable energy sources are typically domestic and distributed, making them more resilient and less vulnerable to geopolitical instability. Renewable energy can also improve access to electricity in remote and rural areas. Many developing countries lack access to reliable energy sources, which can limit economic development and social progress. Renewable energy technologies such as solar panels and wind turbines can be deployed in these areas to provide clean, reliable, and affordable energy.

In addition, green energy can help to conserve natural resources and protect the environment. Fossil fuels are a finite resource, and their extraction and use can have significant environmental impacts, such as oil spills and land degradation. Renewable energy, on the other hand, is derived from sources that are replenished naturally and can be sustainably managed. Finally, green energy can help to promote social and environmental justice. The negative impacts of climate change and environmental degradation are often felt most acutely by marginalized and vulnerable communities. Investing in renewable energy can help to reduce these impacts and create a more equitable and sustainable future for all.

To conclude, as countries around the world are going to transition to a low-carbon future, renewable energy will play a key role in building a more sustainable and resilient world [3].

4. Solar energy

The sun is the only source of energy for all living beings on Earth. Photovoltaic cells work on the principle that solar energy energizes the electrons in the valence band and they are transferred to the conduction band. These electrons are then available for the conduction of electricity. Solar energy can be converted into useful energy for our daily uses directly using various available technologies. These technologies are grouped under a few fundamental categories.

4.1 Solar photovoltaic (PV)

The conversion of solar energy into electrical energy is a relatively newer concept. This was successfully attempted in the year 1975 through the making of photovoltaic cells. These solar photovoltaic cells use solar energy from the sun in order to create electricity *via* the photoelectric effect. These photovoltaic cells have the power to directly convert solar energy into electrical energy. Photovoltaic cells are interconnected with each other to form a photovoltaic module. Photovoltaic modules are the basic structural and functional units of a photovoltaic system. These modules are often connected in series or in parallel in order to get the maximum energy that can be harvested from solar energy. The parallel connections are responsible for raising the energy in the photovoltaic array, while the series connections are used to increase the voltage of the photovoltaic array. These photovoltaic modules often have sizes ranging from 50w to 180w. In direct sunlight, the output goes to a maximum of 180 W. Researchers are looking to increase this number to as much as possible. The higher the total surface area of the photovoltaic cells will be, the higher will be the amount of electrical energy produced for the same intensity of direct sunlight. These modules are connected together in order to form a photovoltaic system and their circuits are usually environmentally protected by lamination [3, 5].

The photovoltaic modules form an integral part of PV systems, as they are positioned in such a way as to enable utilization for various applications. While the main cost of the system is linked to the photovoltaic array, the other components and the overall balance of the system are the main factors determining the reliability, efficiency, and safety of the installation that uses the photovoltaic modules. The sunlight is converted into direct current, and this direct current is then converted into alternating current and adjusted to meet the power requirements of the users as they need. Although photovoltaic cells cannot directly convert solar energy into alternating current, the direct current that is produced has its own functionality. If required the direct current can be converted into alternating current by the use of generators and then be used as per the requirements [5, 6].

4.2 Solar thermal

In the solar thermal process, we produce electrical energy by using the heat energy of the sun radiated through IR radiations. Firstly, solar collectors can be used to absorb solar radiation in order to provide heating of water or space at lower temperatures. Parabolic mirrors can be utilized to concentrate solar radiation on a larger scale, generating higher temperatures. When solar thermal power plants are employed, this solar radiation is focused on a small area to produce high-temperature steam, which is then directed to a turbine generator. This causes the generator to rotate, creating electrical energy. This type of solar energy conversion is mainly used in commercial applications, such as factories [3].

Solar thermal energy can be categorized into two types namely

a. Parabolic trough system

b. Solar tower system

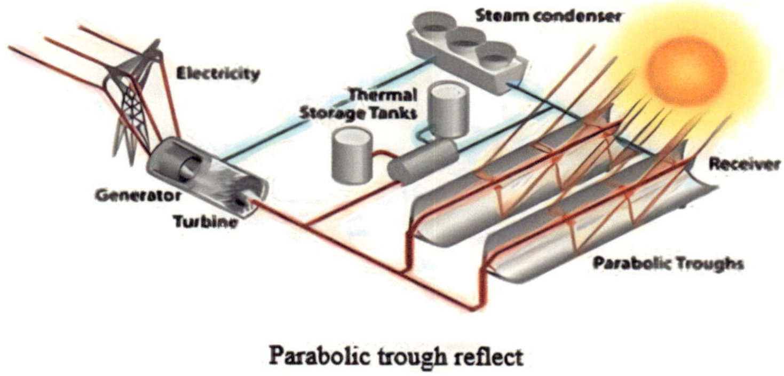

Figure 1.
Parabolic trough system - a parabolic trough system captures sunlight using curved mirrors to heat a fluid in receiver tubes. The heated fluid generates electricity via a turbine or provides heat directly for industrial uses, offering an efficient solar energy solution. (adapted from: https://www.energy.gov/eere/solar/linear-concentrator-system-concentrating-solar-thermal-power-basics).

5. Parabolic trough system

The parabolic system uses mirrors that are parabolic troughs in shape to focus sunlight on a receiver tube. This receiver tube carries a heat transfer fluid, generally, oil or water and this fluid is then heated up and pumped through heat exchangers to produce steam. The liquid can be heated up to as much as 423 K to 623 K as it flows through the receiver and it is then used as a heat source for a power generating system. This heat is then used to run turbine generators to produce electricity *via* the use of a dynamo. In this method, the water or the oil that is used can be reused after cooling them down to a certain extent. This is one of the most economical ways of solar energy conversion. This parabolic trough technology is currently the most proven solar thermal electric technology (**Figure 1**) [3, 7].

6. Solar tower system

Tower systems generally comprise three elements: ground heliostats, a tower, and a central receiver atop the tower. The heliostats capture solar radiation from the sun and direct it to the central receiver. Heliostats rotate in two directions, from east to west and north to south.

Each heliostat is programmed by a computer to follow the sun in order to maximize the total power output. This ensures that the tower traps the maximum amount of solar energy that is available at that point. The first commercial solar tower was built by Abengoa Solar of Spain. It is referred to as PS10 at the Solucar platform in the Spanish province of Seville. It was first inaugurated by the company Abengoa in order to push the production of solar electricity in Spain. The Solucar solar farm currently produces solar electrical energy of almost 3.65 GW. The Solucar solar farm comes second in order of solar production in the world just after Germany. The operation was first started in the month of March 2007 and the process is still under development by the authorities. The Solucar solar farm is the inspiration for multiple similar solar farms all over the world (**Figure 2**) [3, 7].

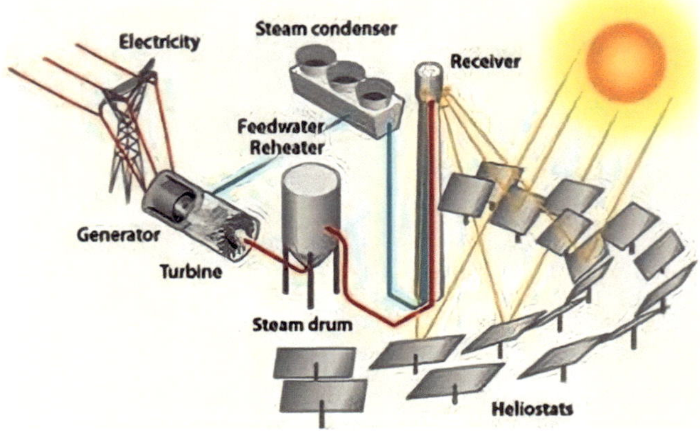

Solar tower system

Figure 2.
Solar tower system - a solar tower system uses an array of mirrors to concentrate sunlight onto a central tower. This intense heat drives turbines to generate electricity, making it a promising solar energy technology with high efficiency and potential for large-scale power generation [adapted from: https://www.energy.gov/eere/solar/linear-concentrator-system-concentrating-solar-thermal-power-basics].

7. Merits of solar energy

The merits of solar energy are many.

1. As in the processing of solar energy, no pollutants are released into the atmosphere so solar energy can be deemed as a clean energy

2. Solar energy can be deemed as the mother of all energy that is present on Earth. Solar energy directly or indirectly gives rise to other forms of energy such as tidal energy and wind energy, all of which are non-polluting.

3. Solar energy may not be available to us all the time but the amount of solar energy that is received by us in the daylight, if stored properly, is more than enough to last us the whole day. To achieve this, photovoltaic cells and batteries are being used all over the world.

4. As per the researcher's data, the Earth receives almost 340 W/m^2 of solar energy on an hourly basis on a non-cloudy day. This vast amount of energy, if harnessed could help us eradicate all of Earth's energy crisis problems [8].

5. There are still areas in the world where there is still no electricity supply. Solar panels and photovoltaic cells give them an opportunity to gain access to electricity which is like a privilege to them. It helps in giving the underprivileged the wonders of electricity. The installation of solar panels and photovoltaic cells is very simple as it does not require wires and receivers. All we need for solar electrical energy is a solar panel and an installation area. Solar panels help cut

energy consumption costs by over 20% over an elongated period of time. This cost-cutting is due to the fact that there is no additional fuel required to run the solar panel other than sunlight [5, 7].

8. Demerits of solar energy

1. Solar panels that are required to harvest solar energy are very expensive. The excess solar energy that is harvested must also be stored for future use, and the available storage solution is also costly. The converters and the inverters that are required to convert the direct current into alternating current and the device to ensure that the supply of current is uninterrupted are not cost-effective. This skyrockets the initial investment of solar panels, which is not affordable by everybody even if it gets cheaper in the long run.

2. Solar panels produce energy in the range of 50 to 180 W. To get a substantial amount of energy, the number of solar panels that are required is huge in number. This creates the problem of area deficiency as a huge number of solar panels require a large amount of area for the maximum output. This restricts people from buying solar panels.

3. Solar panels and photovoltaic cells that are required to convert solar energy to electrical energy are non-recyclable. Solar energy although non-polluting, the damaged photovoltaic plates are treated as garbage and add to the waste of the world.

4. The working and the output of the photovoltaic cells are weather-dependent as they are solely dependent on the amount of solar energy they receive from the sun on a particular day. If the day is cloudy, then it implies that solar cells cannot get the desired amount of solar energy they require to fulfill their needs.

5. The photovoltaic cells, solar energy storage, inverters, and converters, all are high maintenance machines. They require frequent maintenance from trained professionals otherwise they may lead to functional failure which will then mean both loss of money and more wastage for the world as they are not bio-degradable.

6. To get the maximum utilization of money and space, we need a huge number of photovoltaic cells. These photovoltaic cells require the production of single silicon crystal PV systems, which is very technically difficult, energy-intensive, and also very time-consuming [6, 7].

9. Wind energy

Wind energy is a renewable source of energy that has been gaining popularity in recent years due to its clean and sustainable nature. Turbines are used to convert the kinetic energy of wind into electrical energy. This energy production technique does not require any fuel, making it an appropriate option for those looking to reduce their carbon footprint. Wind power is also becoming cost-effective, because of technological advances that have made it more efficient and reliable.

Wind turbines are typically composed of a rotor, which consists of several blades that capture the wind's kinetic energy, and a generator, which converts this energy into electricity. The blades are designed to capture as much wind as possible while minimizing drag and turbulence. The generator then converts this mechanical energy into electrical energy, which can be used for various purposes such as powering homes or businesses [3].

The development of wind power has been driven by government policies that have created a market for renewable energies and by research and development in the field. Advances in control systems, rotor blade profiles, and power electronics have allowed for more efficient use of the wind's kinetic energy. This has resulted in lower costs for consumers while still providing clean and sustainable electricity.

In addition to providing clean electricity, wind power also has other benefits such as reducing air pollution from burning fossil fuels and creating jobs in the renewable energy sector. Wind turbines can also be used to supplement other renewable energies such as solar energy or hydroenergy [9, 10].

Overall, wind power is an attractive option for those looking to reduce their carbon footprint while still providing reliable electricity at reasonable costs. With continued research and development in the field, it is likely that wind power will continue to become more efficient and cost-effective in the future.

In the case of wind energy production, the most important part of this process is the conversion of the kinetic energy of the wind into electrical energy. For this conversion, a wind turbine is used. Over the past few years, average wind turbine ratings have grown almost linearly with current commercial machines rated at 1.5 MW.

Wind turbines are of two types namely horizontal axis wind turbines (HAWT) and vertical axis wind turbines (VAWT). The two varieties of the vertical axis wind turbine (VAWT) are Darrieus and Savonius.

9.1 Horizontal axis wind turbine (HAWT)

A horizontal axis wind turbine, also known as HAWT, is a type of wind turbine that has a rotor blade assembly mounted horizontally on top of a tower. The rotor assembly consists of blades that rotate around a central hub, similar to a propeller. As the wind passes through the rotor blades, it causes the blades to rotate, generating mechanical energy that can be converted into electrical energy. HAWTs are the most common type of wind turbine used for large-scale commercial wind farms. They are typically more efficient than vertical-axis wind turbines and are capable of generating more electricity.

HAWTs can be installed onshore or offshore, and they can be used in a variety of wind conditions, making them a versatile option for renewable energy generation. One advantage of HAWTs is that they can be placed higher off the ground, allowing them to capture more wind energy. They are also easier to maintain and repair than vertical axis wind turbines, as the rotor assembly is mounted at ground level, making it easily accessible. However, HAWTs also have some disadvantages. They can be noisy and can pose a threat to birds and other wildlife. They also require a relatively large amount of space, which can be a challenge in densely populated areas. Overall, horizontal-axis wind turbines are an important part of the renewable energy mix and are likely to continue to play a significant role in meeting the world's energy needs in the coming years [11].

9.2 Vertical axis wind turbine (VAWT)

Vertical axis wind turbines (VAWTs) are innovative wind turbines that have been gaining popularity in recent years due to their unique design and efficiency. Unlike traditional horizontal axis wind turbines, VAWTs have their rotor shaft perpendicular to the ground, which allows them to capture wind.

9.3 Savonius

Unlike the Darrieus VAWT, it is a slow-rotating speed with a high torque machine. It can start at low wind speed, it does not need any external power source to turn on. It does not create so much noise as the Darrieus one. So, it is more environmentally friendly. It does not create any noise pollution. It can work in any wind direction [11].

9.4 Merits of wind energy

Wind energy is a renewable and sustainable source of power that is harnessed from the natural movement of wind. It has emerged as a popular alternative to traditional fossil fuels due to its numerous benefits. In this chapter, we will discuss the merits of using wind energy.

9.4.1 Wind energy is a clean and green source of energy

Wind energy is clean and does not produce any pollutants, such as carbon dioxide, sulfur dioxide, or nitrogen oxide, which can have harmful effects on the environment. Unlike fossil fuels, wind energy does not emit any greenhouse gases, which cause global warming and climate change. The use of wind energy can help reduce the carbon footprint of our energy consumption, making it a sustainable and environmentally friendly solution.

9.4.2 Wind energy is abundant and renewable

Wind energy is abundant and widely available in many regions around the world. As long as the sun shines and the Earth rotates, wind energy will continue to be generated. Wind energy is the fastest-growing renewable energy source in the world, and it is expected to continue to grow in popularity as technology improves and becomes more affordable.

9.4.3 Wind energy is cost-effective

The cost of wind energy has decreased significantly over the past decade, making it increasingly affordable for households, businesses, and governments to invest in. Additionally, once a wind turbine is installed, it has very low operating costs, making it an economically viable solution in the long run.

9.4.4 Wind energy can reduce dependence on fossil fuels

Wind energy can help reduce our dependence on fossil fuels, which are finite resources and have harmful effects on the environment. By using wind energy, we can reduce our reliance on non-renewable sources of energy such as coal, oil, and natural gas, and instead invest in sustainable energy solutions.

9.4.5 Wind energy creates jobs and boosts local economies

The wind energy industry creates jobs in many different areas, including manufacturing, construction, and maintenance. The development and construction of wind farms can also have a positive economic impact on local communities by creating new business opportunities and increasing revenue.

9.4.6 Wind energy is scalable and versatile

Wind energy can be used on a small or large scale, making it a versatile solution for a variety of energy needs. Wind turbines can be installed in remote areas to provide power to off-grid communities, or they can be installed in urban areas to provide power to businesses and households.

9.4.7 Wind energy can improve energy security

By diversifying our sources of energy, we can improve energy security and reduce our vulnerability to supply disruptions or price fluctuations. Wind energy is a reliable and consistent source of power, which can help mitigate the risks associated with energy supply insecurity.

9.4.8 Wind energy is low maintenance and long-lasting

Wind turbines require little maintenance once they are installed, and they have a lifespan of around 25 years. This means that they can provide a stable source of power for many years without significant upkeep costs.

9.4.9 Wind energy is noise-free and non-invasive

Modern wind turbines are designed to be quiet and produce very little noise pollution. Additionally, they are designed to have a minimal impact on the environment and wildlife, making them a non-invasive solution for sustainable energy production.

In conclusion, wind energy is a clean, abundant, cost-effective, and versatile source of power that offers numerous benefits over traditional fossil fuels. By investing in wind energy, we can reduce our carbon footprint, create jobs, boost local economies, and improve energy security. As technology continues to improve, wind energy is poised to become an even more attractive and viable solution for sustainable energy production [12, 13].

9.5 Demerits of wind energy

Wind energy is a popular source of renewable energy, but it has its demerits. In this section, we will discuss some of the disadvantages of wind energy.

1. Intermittency: Wind energy is an intermittent source of energy, which means that it is not available all the time. Wind speeds vary depending on weather conditions, time of day, and season. As a result, wind turbines may not produce energy at a constant rate.

2. Land use: Wind turbines require a lot of land to produce energy on a large scale. This can be a problem in areas where land is deficient. Wind turbines can also have negative impacts on wildlife habitats and migration patterns.

3. Visual pollution: Wind turbines can be seen as an eyesore by some people. They are tall and can be seen from miles away. This can be a problem in areas where there are strict regulations on the use of land and the preservation of natural beauty.

4. Noise pollution: Wind turbines produce noise pollution, which can be a problem for people living in the vicinity of wind farms. The noise can be constant and may affect the quality of life of those living near the turbines.

5. Maintenance costs: Wind turbines require regular maintenance to ensure that they operate at peak efficiency. This can be expensive and time-consuming, especially in offshore wind farms.

6. Bird and bat mortality: Wind turbines can cause bird and bat mortality. These animals can collide with the turbines and suffer fatal injuries. This can be a problem in areas where endangered species are present.

7. Wind energy is weather dependent: As wind energy is dependent on weather, it cannot be relied on to meet the energy needs of a region entirely. Energy storage systems like batteries can help to mitigate this problem, but they are still expensive and not widely used.

In conclusion, wind energy is an excellent source of renewable energy, but it is not without its demerits. The issues of intermittency, land use, visual and noise pollution, maintenance costs, bird and bat mortality, and weather dependency need to be considered when evaluating the potential of wind energy as a sustainable energy source [14, 15].

10. Geothermal energy

The word geothermal energy means energy developed from the inside of the Earth. So, geothermal energy is created by the intrinsic heat of the Earth. The high temperature and pressure in the Earth's core result in the melting of the metals where the high temperature originates from the movement of the tectonic plates and from the radioactive decay of materials.

Geothermal energy can be used in the generation of electricity or directly in space heating, aquaculture, laundry, and industrial processes. Geothermal energy can be found in abundance in hot springs all over the world.

Among these four types, that is, hydrothermal, hot dry rock, magma, and geopressured, only hydrothermal resource is used commercially while the other three resources are still under development. Although hydrothermal energy is being commercially applied in many places, other types of geothermal energy are also used intermittently [3, 16].

10.1 Dry steam technology

Dry steam power plants draw energy from underground resources like steam. The steam is piped directly from underground wells to the power plant where it is directed into a turbine or generator unit which in turn generates electricity. Here, the condensate is usually re-injected into the reservoir or used for cooling.

There are only two known underground resources of stream, The Geysers in Northern California and Yellowstone National Park in Wyoming [17].

10.2 Flash steam technology

Flash steam power plants are the most common type. They use geothermal reservoirs of water with temperatures greater than 455 K. This very hot water flows through wells in the ground under its own pressure. As it flows upward, the pressure decreases and some of the hot water boils or flashes into steam. The steam is then separated from the water and used to power a turbine. Any leftover water and condensed dew are injected back into the reservoir, making this a sustainable resource. The remaining hot water may flash again twice or maybe even three times, at progressively lower pressures and temperatures to obtain more steam. They are the least expensive out of the three geothermal harvesting technologies in both building and operating. They are also the least prone to chemical precipitation problems due to the source of energy being the steam that is formed in flash steam technology [17, 18].

10.3 Binary cycle technology

Binary cycle power plants operate on water at lower temperatures of about 380 to 455 K. These plants use the heat from the hot water to boil a working fluid usually an organic compound with a low boiling point and the working fluid is vaporized in a heat exchanger and used to turn a turbine. The water is then injected back into the ground to be reheated. The water and the working fluid are kept separated during the whole process, so there are little to no air emissions. A higher temperature range provides thermal stability of the working fluid, while the lower temperatures are more feasible in terms of techno-economic and financial factors. Further, the impacts of corrosion and scaling are not apparent at high temperatures, as there is no contact between the power generation equipment and the geo-fluid. These plants are the most versatile of all and the functionality of power is determined by the second cycle [17].

10.4 Merits of geothermal energy

Geothermal energy is a type of renewable energy that is generated by the Earth's internal heat. It has several advantages over other forms of energy, including its availability, sustainability, reliability, and low environmental impact.

1. Availability: Geothermal energy is a renewable energy source that is available at any time. Unlike solar and wind energy, geothermal energy is not dependent on weather conditions or the availability of sunlight or wind. This makes it a reliable source of energy that can be used to meet the demands of a wide range of applications.

2. Sustainability: Geothermal energy is a sustainable energy source because it is continuously replenished by the Earth's internal heat. The amount of heat produced by the Earth's core is estimated to be about 42 million megawatts, which is more than enough to meet the world's energy needs. Unlike fossil fuels, which are finite resources, geothermal energy will not run out as long as the Earth's internal heat is available.

3. Reliability: Geothermal energy is a reliable source of energy because it is not affected by weather conditions. Unlike solar and wind energy, which can be intermittent, geothermal energy can be generated consistently throughout the year. This makes it a reliable source of energy for base load power generation, which is the minimum amount of power required to meet the energy demands of a particular region.

4. Low Environmental Impact: Geothermal energy has a low environmental impact compared to other forms of energy. The process of generating geothermal energy does not produce greenhouse gas emissions, which contribute to climate change. The use of geothermal energy also does not produce harmful pollutants, which can have negative impacts on human health and the environment. In addition, the land use requirements for geothermal power plants are relatively small compared to other types of power plants.

5. Cost-effective: Geothermal energy can be a cost-effective source of energy for power generation. Although the upfront costs of building geothermal power plants can be high, the operating costs are relatively low compared to other types of power plants. This is because the fuel source for geothermal power plants is free, which reduces the overall cost of generating electricity. In addition, the long lifespan of geothermal power plants, which can last up to 30 years or more, reduces the need for frequent replacements or maintenance, which can further reduce the overall cost of generating electricity.

6. Versatility: Geothermal energy can be used for a wide range of applications, including power generation, heating and cooling. In addition, it can be used in combination with other renewable energy sources, such as solar and wind energy, to create hybrid systems that can provide a more stable and reliable source of energy.

7. Local development: The development of geothermal energy can provide benefits to local communities. For example, geothermal power plants can provide a source of revenue for local governments, which can be used to fund local projects and services. In addition, the development of geothermal energy can provide energy security and independence for local communities, reducing their reliance on imported fossil fuels.

8. Reduced energy dependence: The use of geothermal energy can reduce a country's dependence on imported fossil fuels. This can help to improve energy security and reduce the impact of fluctuations in global energy prices. In addition, it can reduce a country's carbon footprint [19, 20].

10.5 Demerits of geothermal energy

However, like all sources of energy, geothermal energy also has its demerits that limit its potential for widespread use.

1. Limited availability: Geothermal energy is not available in all parts of the world. It can only be harnessed in areas where there is enough heat beneath the Earth's surface. These areas are limited to regions with active volcanoes, hot springs, and

geysers. This means that geothermal energy cannot be used as a primary source of energy in areas without these natural features.

2. High upfront costs: The initial cost of building a geothermal power plant is relatively high. The cost of drilling, constructing, and maintaining the wells and the power plant is expensive. The cost of geothermal exploration can also be significant. These high upfront costs can make it difficult for companies and governments to invest in geothermal energy.

3. Environmental impact: Although geothermal energy is a clean source of energy, it can still have a negative impact on the environment. The construction of geothermal power plants can disrupt natural habitats and wildlife. The drilling and injection of water into the Earth can also cause earthquakes and other seismic activities.

4. Limited lifespan of wells: Geothermal wells have a limited lifespan. Over time, the wells can become clogged with minerals and sediment, reducing their efficiency. In some cases, the wells can also become contaminated, making them unusable. This means that geothermal power plants require continuous maintenance and monitoring to ensure their longevity.

5. Energy production fluctuation: Geothermal energy production can be affected by natural events such as earthquakes and volcanic eruptions. These events can disrupt the flow of heat and water, reducing the efficiency of the power plant. This fluctuation in energy production can make it difficult for power plants to meet the electricity demand.

6. Limited scalability: Geothermal power plants have limited scalability due to the limited availability of geothermal resources. This means that the energy produced by geothermal power plants cannot be increased beyond a certain point. This makes it difficult to use geothermal energy as a primary source of energy for large-scale industrial operations.

7. Water usage: Geothermal power plants require a significant amount of water to operate. The water is used to extract heat from the Earth and to generate steam to power turbines. This can put a strain on local water resources, especially in areas where water is deficient in nature.

In conclusion, geothermal energy has its demerits, including limited availability, high upfront costs, environmental impact, limited lifespan of wells, energy production fluctuation, limited scalability, and water usage. These demerits limit the potential for widespread use of geothermal energy. However, with continued technological advancements and innovations, it is possible that some of these demerits can be mitigated, making geothermal energy a more viable source of renewable energy in the future [21].

11. Hydrothermal energy

Hydrothermal energy is classified into two categories. One in which the water is heated up using other sources and the heated water produces vapors that in turn rotates a turbine, and in the process produces energy. Another form of hydrothermal energy is decomposing multiple biodegradable materials underwater at a high temperature and pressure and increasing the carbon and calorific content of the waste that is being decomposed. This process is globally known as hydrothermal carbonization.

11.1 Conventional hydrothermal energy

It is well known that tectonic activities along plate margins are controlled by thermal processes induced by density contrasts and changes in rheology. Vice versa, the tectonic movements could lead to heat accumulation because of differences in the thermal capacities of different strata. This heat trapped in different layers can then be used to heat up the water bodies and produce hydrothermal energy. This hydrothermal energy also helps in understanding the tectonic plate movements, the formation and distribution of geothermal resources, water-rock interactions, and the concentration of ore-forming elements.

Besides the study of tectonics, the heat anomaly in the hydrothermal reservoir might be a result of groundwater movement. In a high-temperature and high-pressure environment, the coupled effect of water and heat serves as the driving forces of groundwater flow. It further dominates the dynamics of deep hydrothermal circulations.

Traditionally, hydrothermal systems only refer to the flow and heat transport processes that redistribute energy and mass in response to circulating fluids such as groundwater and brine (**Figure 3**).

The total geothermal energy in these basins is 2.5 × 1022 J, which is about 260 times the total energy consumed in China in the year 2010. The utilization of hydrothermal resources is undergoing rapid development and strong growth can be further expected. Incidentally, there is still a huge potential for hydrothermal systems that have not been fully exploited. Secondly, the Enhanced Geothermal Systems (EGS) are still in the experimental and pilot stage, which is far from maturity for commercial use. Thirdly, the cost of hydrothermal utilization is generally under 2 ¢/kWh for space heating and 4 ¢/kWh for electric generation, while the lowest possible cost for the utilization of EGS is 5.43 ¢/kWh.

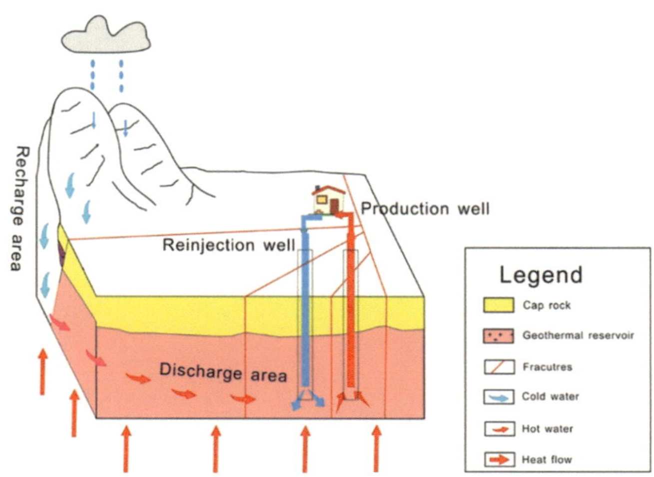

Figure 3.
The conceptual model of reservoir engineering in the hydrothermal systems (adapted from [22]).

We also learned from the case studies that reinjection is absolutely essential for the sustainable utilization of hydrothermal systems. The drawdown of the groundwater table during exploitation is a limiting factor to the long-term, sustainable application of hydrothermal resources.

On the topic of reinjection operation, scientific and technical challenges still remain. For example, the low permeability of sandstone formations and the clogging of reinjection wells often lead to higher costs [22–24].

11.2 Sludge preparation by hydrothermal energy

HTC was applied to stabilize and process sludge collected from septic tanks into hydrochar for practical energy recovery. Hydrothermal carbonization (HTC) is a thermal conversion process that can be used to treat fecal sludge (FS) and convert it into a valuable solid product called "Hydrochar." The process requires short reaction times (1–12 h) at a relatively low-temperature range (180–250°C), with corresponding pressures of up to 30 bar. HTC is applied to stabilize and process FS collected 20 from septic tanks into hydrochar so that pathogens are destroyed and energy recovery is possible.

The study involved using a 1-L high-pressure stainless steel reactor, fitted with a pressure gauge, thermocouple, and gas collection ports. An electric heater with a control panel was used to regulate the reactor temperature and reaction time. The HTC experiment was carried out three times using 350 mL of FS mixed with different additives such as catalysts and biomasses. The process was controlled at a heating rate of 6°C per minute, temperature of 220°C, and reaction time of 5 hours. The pressure inside the reactor was maintained at 30 bar. After each experiment, the reactor was rapidly cooled to ambient temperature using water in a cooling jacket at a rate of about 45°C per minute to stop the reaction. The remaining carbonized FS in the reactor was separated into liquid and solid hydrochar using vacuum filtration (Whatman filter paper, 1.2 μm). The hydrochar was dried in an oven at 105°C for at least 12 hours to remove any remaining moisture. The produced hydrochar was analyzed to determine its characteristics. The mixtures of FS and the selected catalyst or biomass were further tested in the HTC reactor operating at different temperatures and reaction times. The study determined the effects of process parameters such as temperature and reaction time and identified the optimum conditions of the HTC process.

HTC changed the fuel properties of cellulose, along with the results of the ultimate and proximate analyses. The fixed carbon content of cellulose increased from 6.1 to 35.0% in response to HTC at 220°C (**Figure 4**). This result suggests that the cellulose begins to decompose at 220°C. As the fixed carbon content increased during HTC, the calorific value of cellulose increased from 16.5 to 18.9, 23.1, 26.5, and 27.7 MJ/kg at 180, 200, 220, and 280°C, respectively. These increased calorific values can be used to calculate the final effect of HTC over the quality of the biofuel

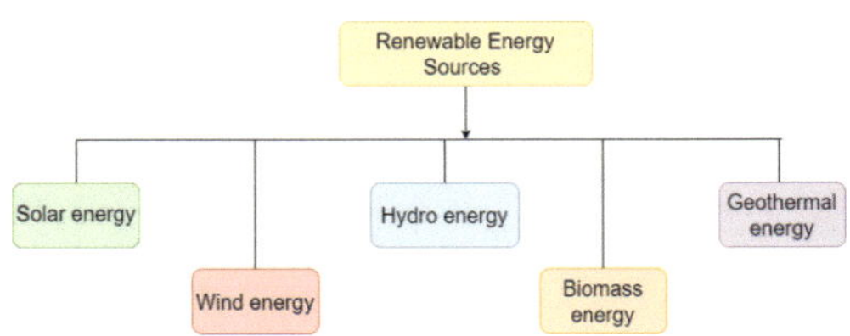

Figure 4.
Classification of different types of renewable energy resources.

(i.e., biochar) obtained from lignocellulosic biomass. This effect can most likely be attributed to the decomposition or pyrolysis of cellulose during HTC reactions owing hydrolysis, chemical dehydration, and decarboxylation reactions.

The HTC process can improve the properties of cellulose in a manner similar to the coalification process. The coalification bands of pure cellulose and its biochar were compared with the coalification bands of various types of coal.

Cellulose is known to have high H/C and O/C ratios, similar to other biomass materials. The H/C and O/C ratios of cellulose decreased with the coalification status between lignite and sub-bituminous coal. This occurred when the cellulose was converted into carbonaceous products by chemical dehydration reactions during HTC [25].

11.3 Merits of hydrothermal energy

Hydrothermal energy is considered a renewable and sustainable energy source as it takes its heat from the Earth's core, meaning that it is essentially inexhaustible. Unlike some other renewable energy sources like wind and solar, hydrothermal energy is available constantly, providing a consistent source of power.

1. It also produces very low levels of greenhouse gas emissions when compared to fossil fuels, making it an environmentally friendly option.

2. Geothermal power plants also have relatively low operating costs, providing stable and predictable energy prices for consumers.

3. Furthermore, they can also provide baseload power, meaning they can meet the constant electricity demand without fluctuations.

11.4 Demerits of hydrothermal energy

On the other hand, geothermal energy resources are limited to regions with suitable geological conditions, meaning its widespread adoption is restricted.

1. Additionally, building a geothermal power plant can be expensive due to the need for exploration and drilling to identify suitable geothermal reservoirs, providing a high upfront cost.

2. Lastly, over time, geothermal reservoirs can become depleted, which reduces energy output and may necessitate the abandonment of a geothermal well. It can limit the long-term sustainability of individual geothermal projects [25].

12. Biomass energy

Biomass energy is a renewable energy source that is produced from organic matter, such as plants, trees, and waste materials. Biomass energy is considered a sustainable alternative to fossil fuels because it is carbon neutral and has the potential to reduce greenhouse gas emissions. In recent years, the use of biomass energy has increased due to concerns about climate change, energy security, and the need to diversify energy sources.

Biomass energy can be produced in several ways, including combustion, gasification, and anaerobic digestion. Combustion is the most common method of biomass energy production, which involves burning biomass to produce heat and electricity. Gasification is a process that converts biomass into a gas, which can be used for heating or electricity generation. Anaerobic digestion is a process that breaks down organic matter in the absence of oxygen, producing biogas that can be used for heating or electricity generation.

Biomass energy can be derived from a variety of sources, including agricultural and forestry residues, municipal solid waste, energy crops, and algae. Agricultural and forestry residues include crop residues, such as corn stalks and wheat straw, and forestry residues, such as tree branches and sawdust. Municipal solid waste includes household garbage, social waste, and industrial waste. Energy crops are crops that are grown specifically for energy production, such as switchgrass and sugarcane. Algae is a type of aquatic plant that can be used to produce biofuels, such as biodiesel and bioethanol.

Biomass energy has several advantages over fossil fuels. One of the primary advantages of biomass energy is that it is a renewable energy source. Unlike fossil fuels, which are finite resources that will eventually run out, biomass can be continually replenished through sustainable practices. Another advantage of biomass energy is that it is carbon neutral. When biomass is burned or converted into energy, it releases carbon dioxide into the atmosphere. However, because the carbon dioxide released during biomass combustion is equal to the carbon dioxide that was absorbed by the plant during its lifetime, biomass energy does not contribute to a total increase in atmospheric carbon dioxide (CO_2) levels [16].

Biomass energy also has the potential to reduce greenhouse gas emissions. By using biomass energy instead of fossil fuels, the amount of carbon dioxide emitted into the atmosphere can be reduced. This is because biomass energy is a low-carbon fuel source that produces fewer emissions than fossil fuels. Additionally, biomass energy can help to reduce waste by using organic materials that might otherwise end up in landfills. Despite its advantages, biomass energy also has some drawbacks. One of the primary drawbacks of biomass energy is that it can be expensive to produce. Biomass energy requires significant upfront investment in equipment and infrastructure, and the cost of biomass feedstocks can be volatile. Additionally, biomass energy can have negative environmental impacts if it is not produced sustainably. For example, clearcutting forests to produce biomass energy can result in habitat destruction and loss of biodiversity.

In conclusion, biomass energy is a renewable energy source that has the potential to reduce greenhouse gas emissions and diversify energy sources. Biomass energy can be produced from a variety of sources, including agricultural and forestry residues, municipal solid waste, energy crops, and algae. While biomass energy has several advantages over fossil fuels, including its renewability and carbon neutrality, it also has some drawbacks, such as high production costs and negative environmental impacts. As research and technology continue to advance, it is likely that biomass energy will play an increasingly important role in the global energy mix [26, 27].

12.1 Merits of biomass energy

i. Biomass is a renewable resource as we can regenerate the lost biomass by growing new crops and trees, making it a sustainable energy source.

ii. As biomass combustion releases CO_2, the same amount of CO_2 is absorbed by the plants used for biomass, forming a carbon-neutral cycle. This, in turn, reduces net greenhouse gas emissions.

iii. Biomass energy is a great alternative to landfills and open burning, as it makes use of agricultural and forestry residues for energy production, thus cutting back on waste.

iv. Since biomass can be sourced locally, it promotes energy independence and reduces reliance on foreign energy sources.

v. We can source biomass energy from a variety of organic materials, providing flexibility in utilization.

vi. With the cultivation and processing of biomass, rural populations can earn jobs and income, promoting energy production in such regions [8, 28].

12.2 Demerits of biomass energy

i. While biomass is considered carbon-neutral over the long term, burning biomass releases CO_2, methane (CH4), and other pollutants into the atmosphere. The carbon neutrality assumption depends on the replanting and growth rates of biomass feedstock, which may not always be guaranteed.

ii. Biomass combustion can produce air pollutants, including particulate matter, nitrogen oxides (NOx), and volatile organic compounds (VOCs), which can have adverse health effects and contribute to smog and air quality issues.

iii. Large-scale biomass production can lead to land-use changes, deforestation, and increased water consumption, potentially impacting ecosystems and water resources.

iv. Biomass energy conversion processes, such as combustion and gasification, are often less energy-efficient than other renewable sources like wind and solar, which can limit their overall environmental benefits.

v. Biomass energy conversion processes, such as combustion and gasification, are often less energy-efficient than other renewable sources like wind and solar, which can limit their overall environmental benefits.

vi. Using land for biomass production can compete with food production and natural habitats, raising ethical and environmental concerns.

vii. Biomass often needs to be transported from rural areas to energy facilities, which can be logistically challenging and energy-intensive, affecting overall efficiency [8, 28].

13. Recent works

There has been a plethora of projects on green energy and chemistry that have been recently adopted by multiple institutions. One of them is the transmission of data using green radio. The specific objective of the Green Radio program is to investigate and create innovative methods for the reduction of the total energy needed to operate a radio access network and to identify appropriate radio architectures that enable such a power reduction. These results clearly show that reducing the power consumption of the base station or access point has to be an important element of this research program. Studies have indicated that the mobile handset power drain per subscriber is much lower than the base station component. Hence, the Green Radio project will mainly focus on base station design issues. This is because the lifetime of a base station is typically 10–15 years, compared to a typical handset being used for 2 years. In addition, the energy costs of a base station are shared between multiple mobile subscribers, leading to a large imbalance in the contribution of embodied energy. From the point of view of devices, significant efforts need to be put into reducing manufacturing energy costs and increasing their lifetime. The Third Generation Partnership Project (3GPP) Long Term Evolution (LTE) system has been chosen as the baseline technology for the research program; its specifications have recently been completed with a view to rolling out networks in the next two to 3 years [29].

Energy can also be gained from the recycling of aircraft's vibration. This can be achieved using the characteristic properties of piezoelectric that convert the vibrational energy into piezoelectricity. Piezoelectricity is the ability of some materials to generate an electric potential in response to applied mechanical stress. This may take the form of a separation of electric charge across the crystal lattice. If the material is not short-circuited, the applied charge induces a voltage across the material [30]. Green chemistry can also be used to produce energy with green technology by Planet-e from living plants and Bacteria. The Plant-Microbial Fuel Cell (P-MFC) is a novel technology that generates electricity in a renewable and likely sustainable way. The P-MFC may mature into a very competitive technology toward other bio-energy systems as it can deliver a net five times higher energy than other systems. The P-MFC concept has several attractive qualities, which can provide a significant breakthrough for sustainable energy production. The plants convert solar energy into organic matter, which is transformed into electricity by electrochemically active bacteria in the fuel cell. Still, the high power output must be maintained, and costs must be reduced to become compatible. Further research and development will reinforce the competitiveness of Europe since P-MFC is worldwide implementable. It has been shown that both the size and type of carbon granules affect the current density of the PMFC. The cathode of the P-MFC can limit the power output [31].

14. Barrier

Implementing green chemistry practices in the field of energy efficiency faces several associated barriers that hinder progress in creating sustainable and environmentally friendly energy solutions. One prominent challenge is the inertia of established industrial processes and technologies, which often resist change due to their long-standing presence and large investments. What's more, there can be economic constraints, as the development and adoption of green chemistry methods may require upfront investments in research and development, making it difficult for some companies to justify the costs. Regulatory barriers, such as outdated or inadequate environmental regulations, can also impede the adoption of greener practices. Furthermore, the lack of awareness and education about green chemistry among stakeholders in the energy sector can slow down its implementation, as it requires a shift in mindset and a commitment to sustainability. Finally, the inherent complexity of energy systems and the need for interdisciplinary collaboration between chemists, engineers, and policymakers pose challenges in developing holistic green chemistry solutions for energy efficiency. Addressing these barriers will be essential to drive the widespread adoption of green chemistry principles in the pursuit of sustainable energy solutions [32].

15. Conclusion

This review highlights the various forms of green energy that we have at our disposal. Though we look at green energy as an alternative source of energy for us in the future, in reality, it might be the only source of energy left for us to explore. The utility of renewable energy and green technology also minimizes the dependency on fossil fuels in the era of global development.

Author details

Suvojit Maity[1], Soham Sarkar[1], Kakoli Dutta[2], Tina De[2] and Ruchira Mukherjee[2]*

1 Department of Electronics and Communication Engineering, Institute of Engineering and Management, University of Engineering and Management, Kolkata, West Bengal, India

2 Department of Basic Sciences and Humanities, Institute of Engineering and Management, University of Engineering and Management, Kolkata, West Bengal, India

*Address all correspondence to: ruchira.mukherjee@iem.edu.in

References

[1] Gielen D, Boshell F, Saygin D, Bazilian MD, Wagner N, Gorini R. The role of renewable energy in the global energy transformation. Energy Strategy Reviews. 2019;**24**:38-50

[2] Farhana N, Gulzar Ahmed M, Asif Iqbal M, Naval Aggarwal N, Biju P, Somayaji A, et al. Fundamental Principles to Address Green Chemistry and Green Engineering for Sustainable Future. London, UK: IntechOpen; 2022. DOI: 10.5772/intechopen.104717

[3] Laxmi Kalyani V, Kumari Dudy M, Pareek S. Green energy: The need of the world. Journal of Management Engineering and Information Technology (JMEIT). 2015;**2**(5):18-26

[4] Kumar CR, Majid JMA. Renewable energy for sustainable development in India: Current status, future prospects, challenges, employment, and investment opportunities. Energy, Sustainability and Society. 2020;**10**(2):1-36

[5] Rabaia MKH, Abdelkareem MA, Sayed ET, Elsaid K, Chae K-J, Wilberforce T, et al. Environmental impacts of solar energy systems: A review. Science of the Total Environment. 2021;**754**:141989:1-19

[6] Patil A, Joseph J, Hawaldar N. A review paper on solar energy-generated electricity. International Journal for Research in Applied Science & Engineering Technology (IJRASET). 2022;**10**(VI):905-2910

[7] Shaikh M, RS, Waghmare SB, Labade SS, Fuke PV, Tekale A. A review paper on electricity generation from solar energy. International Journal for Research in Applied Science & Engineering Technology (IJRASET). 2017;**5**(IX):1884-1889

[8] Joselin Herbert GM, Unni Krishnan A. Quantifying environmental performance of biomass energy. Renewable and Sustainable Energy Reviews. 2016;**59**:292-308

[9] Kumar Y, Ringenberg J, Depuru SS, Devabhaktuni VK, Lee JW, Nikolaidis E, et al. Wind energy: Trends and enabling technologies. Renewable and Sustainable Energy Reviews. 2016:**4**(2)209-224

[10] Kumar A, Khan MZU, Pandey B. Wind energy: A review paper. Gyancity Journal of Engineering and Technology. 2018;**4**(2):29-37

[11] Dauta I, Razlianab ARN, Irwanc YM, Farhanad Z. A study on the wind as renewable Energy in Perlis, Northern Malaysia. Energy Procedia. 2012;**18**:1428-1433

[12] Wang S, Wang S. Impacts of wind energy on environment: A review. Renewable and Sustainable Energy Reviews. 2015;**49**:437-443

[13] Yusuf Alper Kaplan. Overview of wind energy in the world and assessment of current wind energy policies in Turkey. Renewable and Sustainable Energy Reviews. 2015;**43**:562-568

[14] Dai K, Bergot A, Liang C, Xiang W-N, Huang Z. Environmental issues associated with wind energy—A review. Renewable Energy. 2015;**75**:911-921

[15] Premalatha M, Abbasi T, Abbasi SA. Wind energy: Increasing deployment, rising environmental concerns. Renewable and Sustainable Energy Reviews. 2014;**31**:270-288

[16] Ellabban O, Abu-Rub H, Blaabjerg F. Renewable energy resources: Current status, future prospects and their enabling technology. Renewable and Sustainable Energy Reviews. 2014;**39**:748-764

[17] Chijindu Ikechukwu Igwe. Geothermal energy: A review. International Journal of Engineering Research & Technology (IJERT). 2021;**10**(3):655-661

[18] Moya D, Aldás C, Kaparaju P. Geothermal energy: Power plant technology and direct heat applications. Renewable and Sustainable Energy Reviews. 2018;**94**:889-901

[19] Laveet Kumar M, Hossaiz S, Assad MEH, Manoo MU. Technological advancements and challenges of geothermal energy systems: A comprehensive review. Energies. 2022;**15**:9058

[20] Shortall R, Davidsdottir B, Axelsson G. Geothermal energy for sustainable development: A review of sustainability impacts and assessment frameworks. Renewable and Sustainable Energy Reviews. 2015;**44**:391-406

[21] Soltani M, Kashkooli FM, Souri M, Rafiei B, Jabarifar M, Gharali K, et al. Environmental, economic, and social impacts of geothermal energy systems. Renewable and Sustainable Energy Reviews. 2021;**140**:110750

[22] Kong Y, Pang Z, Shao H, Shengbiao H, Kolditz O. Recent studies on hydrothermal systems in China: A review. Geothermal Energy. 2014;**2**:19

[23] Lucian M, Fiori L. Hydrothermal carbonization of waste biomass: Process design. Modeling, Energy Efficiency and Cost Analysis, Energies. 2017;**10**:211

[24] Krysanova K, Krylova A, Zaichenko V. Properties of biochar obtained by hydrothermal carbonization and torrefaction of peat. Fuel. 2019;**256**:115

[25] Koottatep T, Fakkaew K, Tajai N, Pradeep SV, Polprasert C. Sludge stabilization and energy recovery by hydrothermal carbonization process. Renewable Energy. 2016;**99**:978-985

[26] Field CB, Campbell JE, Lobell DB. Biomass energy: The scale of the potential resource. Trends in Ecology and Evolution. 2015;**23**(2):65-72

[27] Mao G, Zou H, Chen G, Huibin D, Zuo J. Past, current and future of biomass energy research: A bibliometric analysis. Renewable and Sustainable Energy Reviews. 2015;**52**:1823-1833

[28] Muhammad Wasif Zafar, Sinha A, Ahmed Z, Qin Q, Zaidi SAH. Effects of biomass energy consumption on environmental quality: The role of education and technology in Asia-Pacific economic cooperation countries. Renewable and Sustainable Energy Reviews. 2021;**142**:110868

[29] Han C et al. Green radio: Radio techniques to enable energy-efficient wireless networks. IEEE Communications Magazine. 2011;**49**(6):46-54

[30] Zhang X, Yang Y, Peng K. Study on energy recycling of aircraft's vibration and noise. In: 2013 IEEE 8th 75 Conference on Industrial Electronics and Applications (ICIEA), Melbourne, VIC. 2013. pp. 617-621

[31] Gurram SPG, Kothapalli NS. A novel electricity generation with green technology by plant-e from living plants and bacteria: A natural solar power

from living power plant. In: 2017 6th International Conference on Computer Applications in Electrical Engineering-Recent Advances (CERA), Roorkee, India. 2017. pp. 146-151

[32] Veleva VR, Cue BW. The role of drivers, barriers, and opportunities of green chemistry adoption in the major world markets. Current Opinion in Green and Sustainable Chemistry. 2019;**19**:30-36